Springer Proceedings in Mathematics & Statistics

Volume 187

Springer Proceedings in Mathematics & Statistics

This book series features volumes composed of selected contributions from workshops and conferences in all areas of current research in mathematics and statistics, including operation research and optimization. In addition to an overall evaluation of the interest, scientific quality, and timeliness of each proposal at the hands of the publisher, individual contributions are all refereed to the high quality standards of leading journals in the field. Thus, this series provides the research community with well-edited, authoritative reports on developments in the most exciting areas of mathematical and statistical research today.

More information about this series at http://www.springer.com/series/10533

Josep Antoni Martín-Fernández
Santiago Thió-Henestrosa
Editors

Compositional Data Analysis

CoDaWork, L'Escala, Spain, June 2015

CoDa

Springer

Editors
Josep Antoni Martín-Fernández
Department of Computer Science, Applied Mathematics and Statistics
University of Girona
Girona
Spain

Santiago Thió-Henestrosa
Department of Computer Science, Applied Mathematics and Statistics
University of Girona
Girona
Spain

ISSN 2194-1009 ISSN 2194-1017 (electronic)
Springer Proceedings in Mathematics & Statistics
ISBN 978-3-319-44810-7 ISBN 978-3-319-44811-4 (eBook)
DOI 10.1007/978-3-319-44811-4

Library of Congress Control Number: 2016948620

Mathematics Subject Classification (2010): 62-XX, 62-07, 62E20, 62F03, 62Hxx, 62Jxx, 62Pxx, 35-XX

Printed on acid-free paper

This Springer imprint is published by Springer Nature
The registered company is Springer International Publishing AG
The registered company address is: Gewerbestrasse 11, 6330 Cham, Switzerland

Foreword

In 1982, Prof. John Aitchison read his well-known article *The Statistical Analysis of Compositional Data* before the prestigious Royal Statistical Society (RSS) in London. Whilst its 21 densely written pages are tremendously stimulating from a scientific point of view, reading the replicas and counter-replicas by prestigious scientists from different areas of knowledge—statistics (eg. D.R. Cox, J.E. Mosimann, D.M. Titterington), geology (eg. R.J. Howart) and finance (eg. G.J. Goodhardt), among others—who critically evaluated the article, is even more so. I have always been of the opinion that almost all of the main issues involved in analysing compositional data (CoDa) and the feasible solutions to them were dealt with during the presentation and subsequent discussions about that historical session of the RSS. Later developments and improvement of the methodology for CoDa carried out by J. Aitchison himself and his continuers have persistently drawn from that presentation

Despite the transcendence of Prof. Aitchison's scientific contribution to CoDa, as often happens in the field of science it remained practically ignored for more than a decade. It was Dr. V. Pawlowsky's innate curiosity and stubborn persistence that made other scientists finally take an interest in this methodology for CoDa, bringing it once again to the fore in discussion forums, initially in the narrow field of geology and later in a wide range of scientific areas of knowledge. CoDaWorks (CDW) illustrates the point well. The first CDW took place at the University of Girona in 2003 with just 28 contributions, while the number of CDWs held to date now totals 6, the last one—with 53 contributions—in L'Escala (Girona) in 2015. CDWs continue to maintain the germinal idea originally conceived: to be a place for theoretic and applied researchers interested in CoDa to exchange ideas. Aside from Prof. J. Aitchison, who despite his advanced years has attended most of the CDWs, other distinguished guest professors invited to participate in the different editions must also be acknowledged: D. Billheimer (Vanderbilt University, USA), G. van der Boogaart (University of Greifswald), A.C. Atkinson (London School of Economics), V. Liebscher (University of Greifswald), J. Bacon-Shone (University of Hong Kong), P. Filzmoser (Vienna University of Technology), M.L. Eaton

(University of Minnesota, USA), C. Glasbey (Biomathematics & Statistics Scotland, UK), C. Reimann (Geological Survey of Norway), R. Tolosana-Delgado (Helmholtz-Institute Freiberg for Resources Technology, Germany), E. Grunsky (University of Waterloo, Canada) and R.S. Kenett (KPA Ltd., Raanana, Israel).

The publication you are holding is a collection of 12 noteworthy contributions to CoDaWork 2015. As you will see, with CoDa as the common denominator they deal with a diverse range of topics from geochemistry to archaeology, and there are also some methodological contributions.

More than 30 years have passed since Prof. J. Aitchison's lone presentation before the scientific collectives of the RSS. This book, wholly dedicated to CoDa, demonstrates that that presentation by the old professor was not given in vain. Let us hope it remains that way for many years to come!

Dr. Carles Barceló-Vidal
Emeritus Professor, University of Girona

Contents

Contributors

C. Barceló-i-Vidal Department of Computer Science, Applied Mathematics and Statistics, Universitat de Girona, Girona, Spain

J. Bergman Department of Statistics, Lund University, Lund, Sweden

Madalyn S. Blondes Eastern Energy Resources Science Center, U.S. Geological Survey, Reston, VA, USA; Department of Geological Sciences, University of Texas at El Paso, El Paso, USA

A. Buccianti Department of Earth Sciences, University of Florence, Firenze, Italy

N. Caciagli Kinross Gold Corporation, Toronto, ON, Canada

I. Dunkl Department of Sedimentology and Environmental Geology, Georg August University, Goettingen, Germany

J.J. Egozcue Department of Civil and Environmental Engineering, Technical University of Catalonia, Barcelona, Spain

M.A. Engle Department of Geological Sciences, University of Texas at El Paso, El Paso, USA

E. Fišerová Department of Mathematical Analysis and Applications of Mathematics, Palacky University, Olomouc, Czech Republic

I. Galván-Femenía Department of Computer Science, Applied Mathematics and Statistics, Universitat de Girona, Girona, Spain

A. Garola Universitat Politècnica de Catalunya (UPC), Barcelona, Spain

N.J. Geboy Eastern Energy Resources Science Center, U.S. Geological Survey, Reston, VA, USA

J. Graffelman Department of Statistics and Operations Research, Universitat Politècnica de Catalunya, Barcelona, Spain

E.C. Grunsky Department of Earth and Environmental Sciences, University of Waterloo, Waterloo, Canada

K. Hron Department of Mathematical Analysis and Applications of Mathematics, Palacky University, Olomouc, Czech Republic

E. Jarauta-Bragulat Department of Enginyeria Civil I Ambiental, Universitat Politècnica de Catalunya, Barcelona, Spain

B.A. Kjarsgaard Geological Survey of Canada, Ottawa, Canada

A. Lindahl Laboratory for Ceramic Research, Department of Geology, Lund University, Lund, Sweden; Department of Anthropology and Archaeology, University of Pretoria, Pretoria, South Africa

G. Mateu-Figueras Department of Computer Science, Applied Mathematics and Statistics, University of Girona, Girona, Spain

T. Monreal-Pawlowsky IZVG, Keighley, UK

G.S. Monti Department of Economics, Management and Statistics, University of Milano-Bicocca, Milano, Italy

B. Nisi CNR-IGG (Institute of Geosciences and Earth Resources), Pisa, Italy

M.I. Ortego Universitat Politècnica de Catalunya (UPC), Barcelona, Spain

R. Ortells The London School of Economics and Political Science (LSE), London, UK

V. Pawlowsky-Glahn Department of Computer Science, Applied Mathematics, and Statistics, University of Girona, Girona, Spain

B. Raco CNR-IGG (Institute of Geosciences and Earth Resources), Pisa, Italy

R. Tolosana-Delgado Helmholtz Zentrum Dresden-Rossendorf, Helmholtz Institute Freiberg for Resource Technology, Freiberg, Germany

K.G. van den Boogaart Helmholtz Zentrum Dresden-Rossendorf, Helmholtz Institute Freiberg for Resource Technology, Freiberg, Germany

Optimising Archaeologic Ceramics h-XRF Analyses

J. Bergman and A. Lindahl

Abstract We present the first results of an experiment which is aimed at ultimately producing recommendations for analysing archaeologic ceramics specimens using handheld XRF analysis devices. In this experiment we study the effects of different measurement durations, different number of measured points and three different types of surface treatments (breakage, polished, grounded) when analysing ceramics specimens, while controlling for nine different types of clay and three different types of temper (no temper, sand, rock), in total almost 1000 analysed points. For each measurement, the proportions of 36 different elements and all other elements are estimated. In the cases with multiple measurements of a specimen, the compositional centre of the measurements is calculated. A complicating issue in the analysis is the large number of parts found to be below detection limit; 13 elements have more than 50 % of the measurements below detection limit and for more than half of those (almost) all measurements are below detection limit. We try nine different strategies for imputing the values. Each estimated elemental composition is compared to a reference estimate using the simplicial distance. The log distances are finally analysed using analysis of variance with main and interaction effects. We find that the different surface treatments have the greatest effect on the distances: grounded specimens yield the most accurate estimates and polished surfaces the least. We also find a significant effect of increasing the number of measured points, but less effect of increasing the duration of the measurements.

J. Bergman (✉)
Department of Statistics, Lund University, Box 743, SE-220 07 Lund, Sweden
e-mail: jakob.bergman@stat.lu.se

A. Lindahl
Laboratory for Ceramic Research, Department of Geology, Lund University, Sölvegatan 12, SE-223 62 Lund, Sweden

A. Lindahl
Department of Anthropology and Archaeology, University of Pretoria, Pretoria, South Africa
e-mail: Anders.Lindahl@geol.lu.se

J.A. Martín-Fernández and S. Thió-Henestrosa (eds.), *Compositional Data Analysis*, Springer Proceedings in Mathematics & Statistics 187,
DOI 10.1007/978-3-319-44811-4_1

Keywords Archaeologic XRF analyses · Archaeometric experiment · Ceramics analysis · Elemental composition analysis · Simplicial distance

1 Introduction

X-ray fluorescence (XRF) analysis, using handheld devices, has gained increased popularity among archaeologist during the recent years, primarily because of its portability and relatively low cost. The analysis produces an estimate of the elemental composition of a specimen. There is, however, not any real knowledge or agreement of how the analyses should be done to obtain the best results. For how long time should a specimen be analysed? How many points on a specimen should be analysed? Which would be preferable, to analyse one point for 4 min or two different points for 2 min each?

To obtain a good measurement one also needs to consider the type of surface that is being analysed. When an archaeologic artefact is encountered, the surface has usually been exposed to various chemical and mechanical interactions with surrounding materials, changing the elemental composition of the surface. Hence the surface might not be representative of the rest of specimen. To overcome this, one could ground the specimen to a fine powder which would mix all parts of the specimen and also remove any effect that large grains might have on the analysis. Another option would be to break off a small piece of the specimen to create a fresh breakage surface gaining access to the interior of the specimen. A third more controllable option would be to remove a part of the surface of the specimen by polishing it with a suitable tool. An important question is how the choice of treatment will affect the analysis. Is one alternative preferable to the others?

In an attempt to answer the questions above, we present some first results of an experiment in which we study the effects of number of points measured, measurement duration and treatment of the surface. The design of the experiment is described in more detail in Sect. 2 and the results of the experiment are presented in Sect. 3.

2 Experimental Design

Nine different, commercially available, clays were purchased. The clays are listed in Table 1. Each clay was partitioned into three parts and different types of temper was applied, i.e. different materials were added to the clays to control for shrinkage as was commonly done in prehistoric and medieval pottery, and still is done today. Sand was added to the first partition, to the second partition crushed rock was added and to the third partition no temper was added. From the in all 27 different clay partitions, small samples were produced resembling potsherds and fired in a modern kiln at 700 °C to resemble the firing of prehistoric and medieval pottery.

We want to investigate the effect of three factors as follows:

Table 1 The nine different clays used in the experiment

No.	Clay type	Description	Max firing temp. (°C)
1	Earthenware	Black	970–1040
2	Earthenware	Red, 25 % grog 0.2 mm	Up to 1220
3	Earthenware	Pale red, mix of natural blue and red clay	950–1000
4	Earthenware	White	1020–1140
5	Earthenware	White, 25 % grog 0–0.5 mm	1000–1280
6	Stoneware	White	1000–1300
7	Earthenware	Red	1000–1150
8	Stoneware	Black, 40 % grog 0–0.5 mm	1220–1260
9	Earthenware	Red, all lime has been washed/removed	950–1000

- The treatment of the potsherd (TREATMENT)
- The duration of the measurement (DURATION)
- The number of measurement points (POINTS).

What treatment of potsherds provides the best estimates? From each of the 27 different types of test-potsherds three replicates were then prepared for analysis in one of the three ways: one test-potsherd was broken to create a breakage surface commonly found in archaeologic ceramic samples, one test-potsherd was polished using a diamond polishing disc to give a 'perfect', smooth surface and one test-potsherd was grounded to a fine power to give a complete mixture of the sample removing any differences between the surface and the interior of the potsherd.

Is there an advantage in measuring a potsherd at several points and calculating the average or does it suffice to only make one measurement? We tried measuring each potsherd at only one point and compared this to measuring it at five different points and then calculating the average (compositional centre) of the five points.

Does longer measurement duration yield better estimates? We used two levels of measurement duration: 60 and 380 s.

Nine clays, three levels of temper, three levels of treatment, two levels of points and two levels of duration yields 324 different combinations. Half of these requiring a single measurement and half requiring five measurements, in total 972 measurements. However, due to the human factor the number of points analysed were in a few cases four or six instead of five, yielding in total 971 measurements. The analysis was done using a portable XRF device providing measurements of 36 elements plus a 'Balance' accounting for all other elements.

2.1 *Measurements Below Detection Limit*

Looking at the measurements we note that a fairly large amount of measurements are below the detection limit (BDL). The number of BDL measurements for each element is given in Table 2. It should be noted that five elements (chlorine, cobalt,

Table 2 The number of measurements below detection limit (BDL) for each element

Si	Ti	Al	Fe	Mn	Mg	Ca	K	P	S	V	Cr
0	0	0	1	432	634	1	0	410	274	24	83
Ni	Cu	Zn	Rb	Sr	Y	Zr	Nb	Ba	Pb	Th	Cl
656	530	25	0	0	1	1	1	11	21	85	965
Co	As	Se	Mo	Ag	Cd	Sn	Sb	W	Au	Bi	U
968	315	971	221	931	922	918	970	658	854	971	422

In total the analysis comprises 971 measurements. Note that all measurements of selenium and bismuth were BDL, and all but one of antimony. Also chlorine, cobalt, silver, cadmium and tin had more than 90 % measurements BDL

selenium, antimony and bismuth) have more than 99 % BDL measurements, and silver, cadmium and tin have all more than 90 % BDL. Furthermore, magnesium, nickel, copper, tungsten and gold have more than 50 % BDL measurements.

It is of course problematic to analyse data with such a large amount of missing measurements. At least two main strategies are conceivable: elements could be excluded or measurements could be imputed. As in all such cases, it becomes a question of retaining information but not altering the data too much. Imputing data is of course not a problem when only a limited number of measurements are imputed, but one can question the reasonableness of imputing almost all values. We have chosen to try different ways of excluding and imputing in order to compare the effects of the different strategies. The procedure was done in two steps. First three data sets were created removing all elements with more than 50, 90 and 99 % BDL, respectively. In those few cases where elements with observed measurements where excluded, the observed measurements were added to the balance. From a theoretical point of view, it may be noted that this amalgamation is not subcompositionally coherent; however, it was chosen in order to retain as much information as possible. (A subcompositional approach was also tried and the differences were negligible.)

Second, three imputation schemes were implemented to each data set. A non-parametric imputation with 0.65 of the detection limit, a non-parametric multiplicative Kaplan–Meier smoothing spline, and a model-based lognormal with fixed imputation values [3]. All imputations were done using the functions `multRepl`, `multKM` and `multLN` in the R package *zCompositions* [4]. Looking at the average composition (compositional centre) for each combination of clay and temper indicated that there were large differences between the different clays, e.g. the amount of iron, manganese and calcium, but no evident differences due to different temper. For this reason the Kaplan–Meier and the lognormal imputations were done separately for each clay but for all tempers as to retain a reasonable sample size. It should be noted that normally one would not impute parts with more than 50 % BDL values.

To provide some sort of comparisons of the effects of the imputation, the first two principal components of the nine data sets are plotted in Fig. 1a and the third and fourth are plotted in Fig. 1b. The four components account for 81–89 % of the variation in the data sets. The plots in Fig. 1 seem to indicate that overall pattern is the

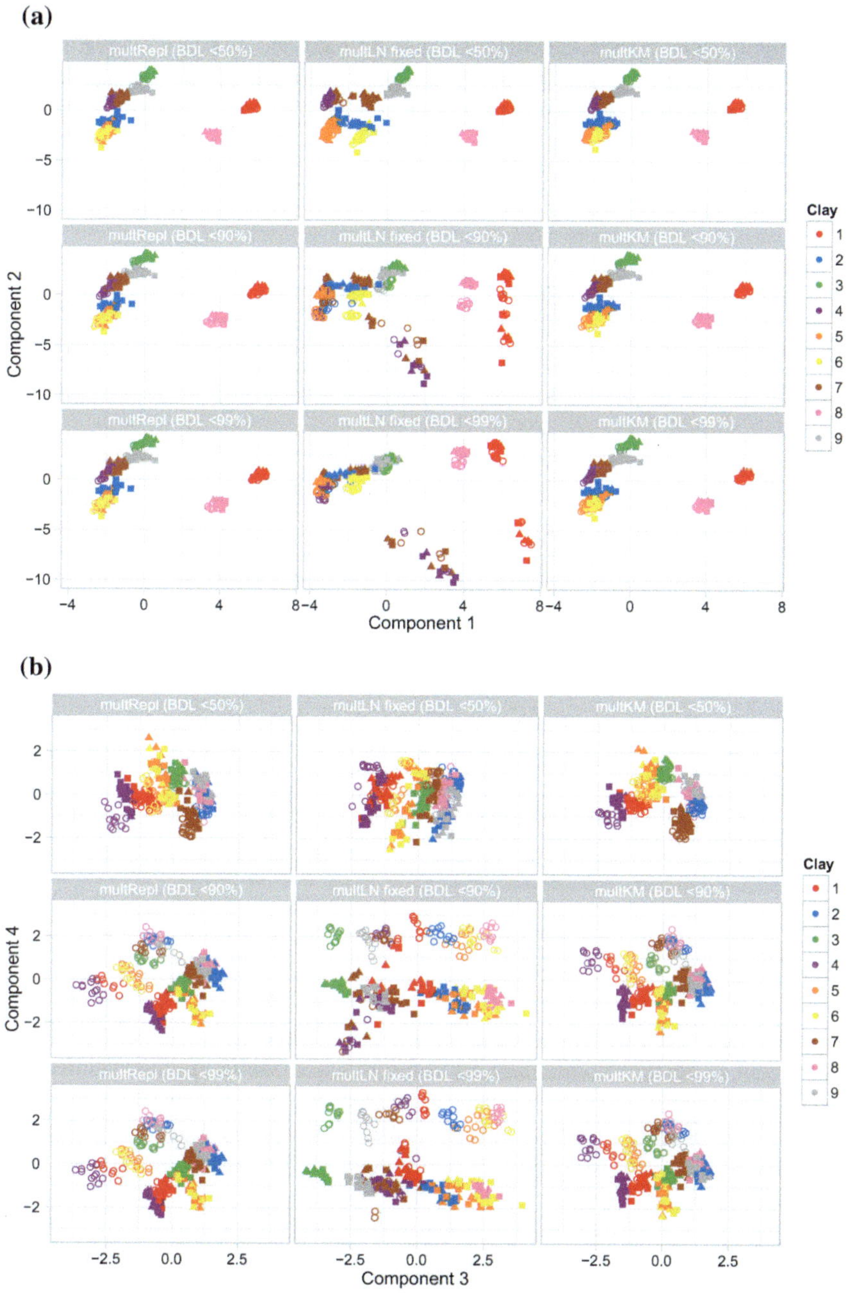

Fig. 1 The (**a**) first and second and (**b**) third and fourth principal components from the compositional PCA analysis of the nine different data sets, with different colours for the different clays (see *colourbar* on the right) and different symbols for the different treatments: breakage surface (■), polished surface (▲), grounded (○)

same in all plots; especially the non-parametric and the Kaplan–Meier imputations yield quite similar patterns. The log-normal imputation, however, seems to have induced more variability, particularly when only elements with less than 90 % or 99 % BDL are retained. In summary, the largest differences between the data sets seem to be between only retaining elements with less than 50 % BDL and keeping more elements, and between the lognormal imputation and the other two imputations.

2.2 *Assessing the Accuracy of the Measurements*

In order to assess the accuracy of the measurement, a reference (or 'truth') is needed. During the summer of 2013, we made an agreement with a colleague who had access to analytical equipment of greater accuracy to analyse the 27 different clay-temper combinations and provide reference measurements. To date we have not received the results, but hopefully they will arrive in the near future. However, we still needed a reference, so we decided to use the results we had. It was deemed that the grounded samples with the longer duration would provide the best estimates. So, for every clay-temper combination the centre composition [2]

$$\mathscr{C}\left(g(x_{11},\ \dots,\ x_{n1}),\ \dots,\ g(x_{1D},\ \dots,\ x_{nD})\right)$$

of the $1+5$ measurements at 380 s was calculated. Here the closure operation is denoted $\mathscr{C}(\mathbf{x}) = (x_1,\ \dots,\ x_D)/\sum_{i=1}^{D} x_i$ and $g(x_1,\ \dots,\ x_n) = (x_1 \dots x_n)^{1/n}$ denotes the geometric mean. This was done separately for the nine different imputation schemes, thus obtaining nine different reference sets each consisting 27 reference compositions.

3 Analysis

For each combination of clay, temper, treatment, number of measured points and measurement duration, we calculate the compositional centre of the measurements. Thus, for one measured point we keep that measurement and for five points we calculate the centre of the five measurements. This is repeated for all imputation schemes resulting in nine data sets of 324 compositional estimates. For each estimate we calculate the simplicial distance [1, p. 64]

$$d_S(\mathbf{x},\ \mathbf{y}) = \sqrt{\sum_{i=1}^{D}\left(\log\frac{x_i}{g(\mathbf{x})} - \log\frac{y_i}{g(\mathbf{y})}\right)},$$

where $g(\cdot)$ denotes the geometric mean, $\mathbf{x}$ denotes the compositional estimate, and $\mathbf{y}$ denotes the corresponding reference composition, i.e. the composition of that com-

Table 3 The results of the analysis of variance for lognormal imputation retaining only elements with less than 50 % BDL

Factor	Df	Sum Sq	Mean Sq	F value	p-value[a]
Clay	8	4.498	0.562	0.9202	0.4999
Temper	2	1.611	0.806	1.3188	0.2690
TREATMENT	2	272.368	136.184	222.9032	0.0000***
DURATION	1	1.890	1.890	3.0930	0.0796
POINTS	1	10.272	10.272	16.8131	0.0001***
TREATMENT:DURATION	2	19.062	9.531	15.6004	0.0000***
TREATMENT:POINTS	2	9.662	4.831	7.9073	0.0004***
DURATION:POINTS	1	3.482	3.482	5.6994	0.0176*
Residuals	304	185.731	0.611		

[a] Significance codes: *0.05 **0.01 ***0.001

The results for the other imputation schemes are similar. The treatment (breakage, polished or grounded) is the most significant factor and the duration of measurements the least significant. The main difference between the imputation schemes is that duration is significant when more elements are retained, but not when only elements with less than 50 % BDL are retained

bination of clay and temper, resulting in nine sets of 324 distances. The calculations are done using the R package *compositions* [5].

The logarithm of the distances are analysed using analysis of variance. (The logarithm of distances are used as the distances are only positive and are expected to have a skew distribution. The decision is further strengthened by the fact that Box-Cox transformations indicate an optimal $\lambda \approx 0.2$, i.e. fairly close to 0.) We model the effect of different treatments, number of points and measurement duration including all two-way interactions, controlling for different clays and temper.

In Table 3 we present results of the analysis of variance for one of the imputation schemes, the lognormal with less than 50 % BDL. It can be noted that the clearly most significant factor is the treatment, i.e. if the measurement was done on a breakage surface, a polished surface, or on the grounded sample. The least significant factor is the duration of the measurement. The results are similar for the other imputation schemes. The only difference is that the duration becomes significant when the number of elements is increased.

To get an idea of how the distances differ for different factor levels, we calculate the estimated expected log distances for the various combinations of treatment, number of points and measurement duration, i.e. the predicted value for each combination of factor levels. Since the expected value also depends on clay and temper we present how the values differ from the baseline of one measured point for 60 s on a breakage surface, the effect of the clay and temper is thus cancelled out. The expected values are presented in Table 4. The shortest distances are found when the samples are grounded and the longest for the polished surfaces, with the breakage surfaces in-between. An interesting observation is that, whereas the accuracy of the measurements are improved with increased measurement duration for the grounded sample, the accuracy deteriorates with increased measurement duration for polished

Table 4 Differences in expected value of the log distances for the various combinations of treatment, number of points and measurement duration compared to the baseline of one measured point for 60 s on a breakage surface

Points	Duration (s)	Treatment		
		Breakage	Polished	Grounded
1	60	0	0.2700	−0.9116
	380	0.1483	0.8663	−1.4924
5	60	0.0422	0.4150	−1.5452
	380	−0.2242	0.5967	−2.5407

These estimates come from the analysis of the lognormal imputation data set retaining only elements with less than 50 % BDL, but are similar for the other imputation schemes

surfaces. For increased number of measurements, the accuracy is improved for 380 s duration for both breakage and polished surfaces but deteriorated for the shorter duration.

Figure 2 shows normal QQ plots of the residuals for each of the nine different analyses (imputation schemes). The plots indicate that the residuals have a slightly skewed distribution possibly violating the normality assumption.

Apparently, the treatment has the greatest impact on the distances, and especially whether or not the sample was grounded. Since grounded samples were used to create the elemental reference compositions, we rerun the analyses without the grounded samples, i.e. with only breakage and polished surfaces. In all the nine data sets treatment and the interaction between treatment and duration are the only significant effects. As the results are similar for all data sets, we provide as an illustration in Table 5 the differences in expected values of the log distances for the various

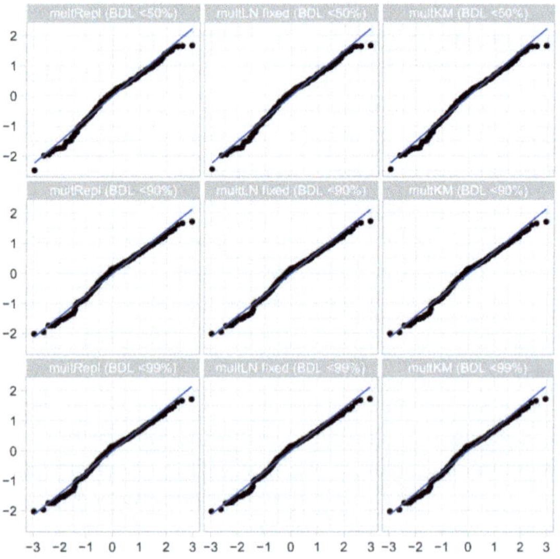

Fig. 2 Normal QQ plots of the residuals for the nine different imputation schemes

Table 5 Differences in expected value of the log distances for the various combinations of treatment, number of points and measurement duration

Points	Duration (s)	Treatment	
		Breakage	Polished
1	60	0	0.2700
	380	−0.1292	0.5888
5	60	−0.2353	0.1376
	380	−0.2242	0.5967

The changes are compared to the baseline of one measured point for 60 s on a breakage surface. The estimates come from the analysis of the lognormal imputation data set retaining only elements with less than 50 % BDL

combinations of treatment, number of points and measurement duration for the same data set as above, i.e. the lognormal imputation with less than 50 % BDL. We note that the breakage surface still provide shorter distances than the polished surface. An interesting observation is that for breakage surfaces, the mean distance decreases when either of the number of points or the measurement duration is increased, but when both are increased not much is gained.

4 Conclusions and Future Research

We have investigated how the accuracy of the elemental composition analysis of ceramics specimens, using a handheld XRF analysis device, is affected by different types of surface treatments, different number of points measured and different measurement durations. Our prior belief was that the more points and the longer the duration, the better the accuracy. We also believed that a grounded specimen would produce the most accurate measurements. These suppositions are confirmed by the analyses.

We can conclude that grounding the specimen is the most important factor in obtaining an accurate measurement. We are actually surprised by the large differences between the three surface treatments. The polished surface should provide an optimal surface for the XRF device, but turns out to yield the worst results. A possible explanation could be that the samples have been polluted by substances from the polishing disc, even though this seems unlikely. Another possible explanation could be that with the polished surface the large grains of, e.g. temper are clearly visible, allowing the analyst to avoid them. This is not possible with a breakage surface. It is, however, a gratifying result that the breakage surface does so well. Grounding and, to a slightly lesser extent, polishing are both destructive treatments, which are often not an option for an archaeologist. One reason for the popularity of the handheld XRF device is that it can be used on artefacts in e.g. museums without damaging or even removing them. Grounding the specimen is thus an optimal but perhaps more theoretical alternative.

At least for breakage surfaces and grounded samples, the most accurate measurements are obtained when using both five points and a measurement duration of 380 s per point. (This is not the case for polished surfaces, which is rather puzzling.) However, from a practitioner's point of view, it should be noted that five measurements each during 380 s means that the total time of the analysis is more than 30 min, not counting the time setting up the equipment, preparing the specimen and moving the specimen between the analyses. Since time is limited (for most of us), the question becomes whether it is preferable to analyse a specimen at only one point for 380 s or at five points for 60 s each? In each case the total time of analysis is about 6 min. Our findings clearly show that measuring five points during 60 s yields more accurate estimates than one point during 380 s. It should be noted though, that for breakage and polished surfaces a single measurement of 60 s gives more accurate estimate when the grounded samples are included in the analysis.

This has been a first report from an ongoing experiment. It is of course highly unsatisfactory to use the same measurements that are analysed to estimate the reference values. We are therefore eagerly looking forward to get new independently estimated reference values. In this paper an analysis of 971 measurements was presented. In total, the experiment to date consists of more than 1800 measurements and more than 139 h of XRF device running time. In order to obtain a balanced experiment, about half of the measurements were excluded, as not all combinations of the factors are currently measured. Thus it remains to complete the measurement sequence. This will hopefully also allow us to identify any threshold values in number of points and measurement duration: How much is gained in accuracy when increasing the number of points from one to three compared to increasing the number of points from three to five? Is there an optimal combination of number of points and measurement duration? It also remains as future research to investigate why the breakage and polished surfaces yielded more accurate measurements when measured only once for 60 s, than compared to five measurement for 60 s and to one measurement for 380 s. Is there a reason for this, or is it some sort of artefact of the extremely strong treatment effect? A final issue is that even though in theory all XRF devices should yield similar estimates, it remains to confirm the results using different XRF devices.

Our conclusions in this experiment so far are that an archaeologist intending to do an elemental analysis of a ceramic specimen using a hand held XRF analysis device should ground the specimen if possible, and if not possible find a fresh breakage surface, and analyse the specimen at five points for 60 s each or, time permitting, for 380 s each and finally calculate the compositional centre of the measurements.

References

1. Aitchison, J.: Principal component analysis of compositional data. Biometrika **70**, 57–65 (1983)
2. Aitchison, J.: Measures of location of compositional data sets. Math. Geol. **21**, 787–790 (1989)

3. Palarea-Albaladejo, J., Martín-Fernández, J.A.: Values below detection limit in compositional chemical data. Anal. Chim. Acta. **764**, 32–43 (2013)
4. Palarea-Albaladejo, J., Martín-Fernández, J.A.: zcompositions: imputation of zeros and nondetects in compositional data sets. R package version 1.0.3, (2014)
5. van den Boogaart, K.G., Tolosana, R., Bren, M.: Compositions: compositional data analysis, R package version 1.40-1 (2014)

A Practical Guide to the Use of Major Elements, Trace Elements, and Isotopes in Compositional Data Analysis: Applications for Deep Formation Brine Geochemistry

M.S. Blondes, M.A. Engle and N.J. Geboy

Abstract In the geosciences, isotopic ratios and trace element concentrations are often used along with major element concentrations to help determine sources of and processes affecting geochemical variation. Compositional Data Analysis (CoDA) is a set of tools, generally attuned to major element data, concerned with the proper statistical treatment and removal of spurious correlations from compositional data. Though recent insights have been made on the incorporation of trace elements and stable isotope ratios to CoDA, this study provides a general approach to thinking about how radiogenic isotopes, stable isotopes, and trace elements fit with major elements in the CoDA framework. In the present study, we use multiple data sets of deep formation brines and compare traditional mixing models to their CoDA counterparts to examine fluid movement between reservoirs. Concentrations of individual isotopes are calculated using isotopic ratios and global mean isotopic abundances. One key result is that isotope parts (e.g. ^{18}O, ^{17}O, ^{16}O, ^{2}H, ^{1}H, ^{87}Sr, ^{86}Sr) can simply be modelled by the major element concentration (H_2O, Sr) in a clr-biplot as they are perfectly dependent. Another important result is that an ilr transformation of radiogenic isotope parts (e.g. ^{86}Sr and ^{87}Sr in $^{87}Sr/^{86}Sr$) and trace elements can, like stable isotopes in delta notation, be treated as a linear function of the isotopic ratio or trace element concentration, scaled only by a constant. This implies that there are multiple situations in which an ilr transformation provides little additional insight for the analysis of trends: (1) any two parts with low log ratio variance (e.g. an isotope ratio), no matter their concentrations in the solution, (2) any low concentration parts (trace elements) or a ratio of a trace to a major element, no matter the variance of the elements, and (3) large positive ratios (major/trace) over a restricted range of variance. Similarly, a multivariate ilr transformation of a large data set with many parts will also be a simple perturbation if the balances are evenly split between parts.

M.S. Blondes (✉) · M.A. Engle · N.J. Geboy
Eastern Energy Resources Science Center, U.S. Geological Survey,
Sunrise Valley Dr., MS 956, Reston, VA 12201, USA
e-mail: mblondes@usgs.gov

M.A. Engle
Department of Geological Sciences, University of Texas at El Paso,
500 W. University Ave., El Paso 79968, USA

J.A. Martín-Fernández and S. Thió-Henestrosa (eds.), *Compositional Data Analysis*, Springer Proceedings in Mathematics & Statistics 187,
DOI 10.1007/978-3-319-44811-4_2

CoDA transformations, however, even if they do not provide new insight in some specific cases, will provide consistent interpretations for all types of data.

Keywords Compositional data analysis · Isotipes · Trace elements · Brines · Produced waters

1 Introduction

In the geosciences, variations in stable isotope ratios, radiogenic isotope ratios, and trace element concentrations are often combined with variations in major element concentration to interpret geologic processes. Radiogenic isotope systems are those in which one or more isotopes are created from the radioactive decay of a parent isotope. Ratios of isotopes from radiogenic systems (e.g. $^{87}Sr/^{86}Sr$) in minerals are a product of initial isotopic abundances, partitioning of elements during crystallization, and subsequent radioactive decay over time. Particular minerals and rock types often have a specific radiogenic isotopic signature that may be imparted onto waters that interact with them. Radiogenic isotopes therefore are often used to determine the origin or pathways of fluids from rocks with a known or assumed isotopic composition or to determine mixing relationships between multiple sources. Stable isotopes of a given element (e.g. ^{16}O, ^{17}O, ^{18}O or ^{1}H, ^{2}H), on the other hand, partition from one another during numerous geologic and biologic processes as a function of differences in the mass and bond energy of the various isotopes (e.g. Faure and Mensing [10]). Variations in stable isotopes can record evidence of specific processes (e.g. evaporation) or mixing between fluids that previously underwent different geologic processes over time (e.g. mixing between isolated basin reservoir brines and fresh meteoric water). Trace element concentrations are also particularly useful for geologic interpretation and, like isotopes, can be used to identify the origin of or processes that geological media have experienced. This is partly because trace element concentration variance can be orders of magnitude larger than the major constituents of a system, making them more sensitive and indicative of physical, chemical, and biological processes (e.g. Goldschmidt [13]).

It has long been understood that compositional data, which include geochemical data, are constrained by closure and subject to spurious correlations without proper treatment [1, 19]. Geologists and statisticians have developed a number of approaches over the years to remove closure, beginning with using concentration ratios to examine only relative information between ions (e.g. Chayes [4]). Another common current method for avoiding the effects of closure has been the use of trace element, rather than major element, concentrations, under the assumption that trace elements are dilute solutions that obey Henry's Law, whereby the trace element activity is proportional to its concentration (e.g. Irving [16]). This proportionality implies that, thermodynamically, trace elements behave independently of major elements like gases in an ideal solution, which has been interpreted to mean trace elements are not subject to the effects of closure. It can be argued, however, that non-interaction between trace and major elements does not necessarily imply compositionally open data. Other approaches use normalized log ratios like the additive log ratio (alr) and centered log ratio (clr) to remove compositional closure and transform the data to a Euclidean space to allow for proper treatment of many statistical tests (e.g. Aitchison [1]). A more recent development is the ilr transformation [9], which has a number of properties (orthogonality, subcompositional coherence, etc.) that make almost any statistical analysis that relies on a Euclidean distance possible (e.g. linear regression and cluster analysis). However, the ilr transformation requires some experience to gain an intuitive sense for interpretation, particularly for exploratory analysis of trace elements and isotopes in geologic data and is not yet common in most geochemical publications. More recent studies [20, 24] addressed the incorporation of stable isotopes with major element compositional data into Compositional Data Analysis (CoDA), an important step toward including the interpretive tools that geologists use.

However, when considering the proper statistical treatment for geochemical data, there is a dilemma: on the one hand, an ilr transformation to orthonormal coordinates will transform the data to the appropriate scale where Euclidean distance can be used, but the data have often been transformed to a point that detailed processes may be difficult to interpret. On the other hand, using simple trace elements and isotope ratios may be easily interpretable to the geologist, but it is not always clear whether these interpretations are mathematically valid. Our goal with this paper is twofold: (1) to build on previous CoDA isotope work to present a general approach to incorporating radiogenic isotopes and trace elements along with stable isotopes and major element chemistry, and (2) to show under what conditions log ratio transformations provide a more robust treatment of geochemical data and under what conditions simple isotopic ratios and trace elements are nearly as valid in terms of following the rules of Euclidean geometry.

Our example for the use of isotopes and trace elements in CoDA are data from deep basin brines, including waters sampled from mine shafts and produced waters. Produced waters are the waters co-generated with hydrocarbons during oil and gas development. Produced waters may include portions of water originally present in geologic formations (formation water) prior to oil and gas production, waters injected for well stimulation and hydraulic fracturing, and water condensing from the gas phase. In many cases the salinity of deep basin brines can exceed that of bulk modern seawater ($\sim$35 g/L) by several times. Previous work has shown that the large range in salinity in produced waters can induce obvious effects on the related composition data, such as spurious correlation [6–8]. Analyzing brines can help researchers understand basin scale hydrogeology and the transport of injected fluids. Isotopic data are particularly useful in interpreting these types of processes because they can demonstrate whether fluids in different reservoirs have mixed, and therefore whether injected fluids have been transported from one reservoir to another. We use two comprehensive chemical and isotopic datasets of formation water to address CoDA of isotopes: (1) formation waters from potash mine shafts in Saskatchewan, Canada [17] and (2) formation brines from a Permian salt dome in the North German Basin [18]. We focus on two stable isotope systems (O and H) and one radiogenic isotope system ($^{87}Sr/^{86}Sr$). Trace elements are represented by Sr and B concentrations, as well as the calculated concentrations of individual low abundance isotopes (e.g. ^{87}Sr, ^{1}H).

2 Treatment of Isotopic Data for Compositional Data Analysis

Whereas major cation (e.g. Ca, Na) and anion (e.g. SO_4, Cl) concentrations are measured as parts of a whole fluid, many isotopic ratios are directly measured as proportions (e.g. $^{18}O/^{16}O$) using a mass spectrometer. In traditional approaches, isotopic ratios are included together with concentrations during data analysis. However, in order to combine isotopic ratios and concentration data in CoDA, particularly for multivariate analyses, it is important to have a comprehensive understanding of how to properly transform all relevant data into the Euclidean geometry of real space.

Tolosana-Delgado et al. [24] showed that chemical concentrations (e.g. SO_4) can be split into separate parts (e.g. $^{34}SO_4$, $^{32}SO_4$) based on their isotopic proportions. They also showed that an ilr transformation of a stable isotope ratio is proportional to values expressed in classical delta notation (per mil (‰) relative to a standard) and therefore the delta notation can simply be scaled to compare isotopic data and major element concentrations simultaneously. Puig et al. [20] applied this approach to discriminant analysis of groundwaters by separating clr transformed compositional data from raw isotopic data and scaled both to have equal variances. This was necessary because the log ratio variance of isotopic ratios is typically much smaller than the log ratio variance of the compositional ionic data.

Display of stable isotopic ratios on a manageable scale is often done using delta notation, in which the isotope ratio in question is normalized to a standard and multiplied by a factor of 1000. For example:

$$\delta^{18}O = 1000 \times \frac{\left(\frac{^{18}O}{^{16}O}\right)_{sample} - \left(\frac{^{18}O}{^{16}O}\right)_{standard}}{\left(\frac{^{18}O}{^{16}O}\right)_{standard}} \tag{1}$$

and

$$\delta D = \delta^{2}H = 1000 \times \frac{\left(\frac{^{2}H}{^{1}H}\right)_{sample} - \left(\frac{^{2}H}{^{1}H}\right)_{standard}}{\left(\frac{^{2}H}{^{1}H}\right)_{standard}} \tag{2}$$

In the case of high salinity samples, such as many formation waters, large differences exist between isotopic ratios of H and O depending on whether they were measured on a concentration (mass) basis or an activity basis. For correct conversion of the data into ilr or clr transformed results, and for comparison with concentration data, use of O and H isotopic data on a concentration basis is critical. Results presented in an activity basis can be converted to a concentration basis using the methods of Sofer and Gat [22, 23].

By comparison, radiogenic isotopic data are conventionally presented as directly measured concentration ratios of the radiogenic to a stable isotope or between two radiogenic isotopes (e.g. Faure and Mensing [10]). For instance, in the case of strontium, radiogenic ^{87}Sr (which is derived from the decay of ^{87}Rb) is typically normalized to ^{86}Sr, one of the four stable Sr isotopes. For radium, different variations exist, but commonly ^{228}Ra (part of the thorium radioactive decay chain) activity is normalized to ^{226}Ra (part of the uranium radioactive decay chain) activity. Often radiogenic isotopic data are not normalized to any standard, although there are exceptions (Sr data can be normalized by modern seawater and reported in epsilon notation).

As Tolosana-Delgado et al. [24] point out, isotopic data in delta notation are proportional to log ratios, but multiplied by a different scaling factor. Unlike SO_4 concentration data that can be split into proportions by its isotopic ratio, O and H isotopes of water must be treated differently because the concentration of pure H_2O in aqueous samples is not usually measured. Further, to perform certain CoDA transformations it is useful to treat each individual isotope as a separate concentration (e.g. ^{16}O, ^{17}O, ^{18}O, ^{1}H, ^{2}H, ^{86}Sr, ^{87}Sr, etc.).

All concentration data in this study are presented in units of mg/kg solution (ppm). The H_2O concentrations were calculated as the difference between the samples' density and total dissolved solids (TDS) concentration, where TDS is equivalent to the sum of all dissolved ions in the fluid. For samples where density was not reported, it was estimated using its relationship with TDS known from brine samples from the Permian Basin of Texas and New Mexico. Water concentrations are therefore perfectly dependent on the concentration of the other ions and inversely correlated with TDS. Based on stoichiometric relationships, H_2O concentrations were converted to concentrations of O and H in water in the solution. For O and H isotopic data in delta notation, $^{18}O/^{16}O$ and $^{2}H/^{1}H$ ratios were determined from Eqs. 1–2, using the known composition of Vienna Standard Mean Ocean Water (VSMOW). Assuming the $^{17}O/^{16}O$ ratio is constant and that of the average global abundances, and assuming ^{3}H is negligible, the concentrations of ^{16}O, ^{17}O, ^{18}O, ^{1}H, and ^{2}H were individually calculated. For Sr isotopes, the Sr concentration in mg/kg solution is split into 4 isotopic concentrations (^{84}Sr, ^{86}Sr, ^{87}Sr, and ^{88}Sr) using the average global abundances for ^{84}Sr, ^{86}Sr, and ^{88}Sr, and the measured $^{87}Sr/^{86}Sr$ ratio.

3 Interpretation of Isotopic Data Using Ilr Transformed Subcompositions

Ilr transformation of subcompositions can be utilized to interpret brine geochemistry. Engle and Rowan [8] showed that not only can ilr transformations of the Na–Cl–Br system reveal halite dissolution and seawater evaporation trends common in certain environments that are apparent in traditional ratio plots, but they also more clearly

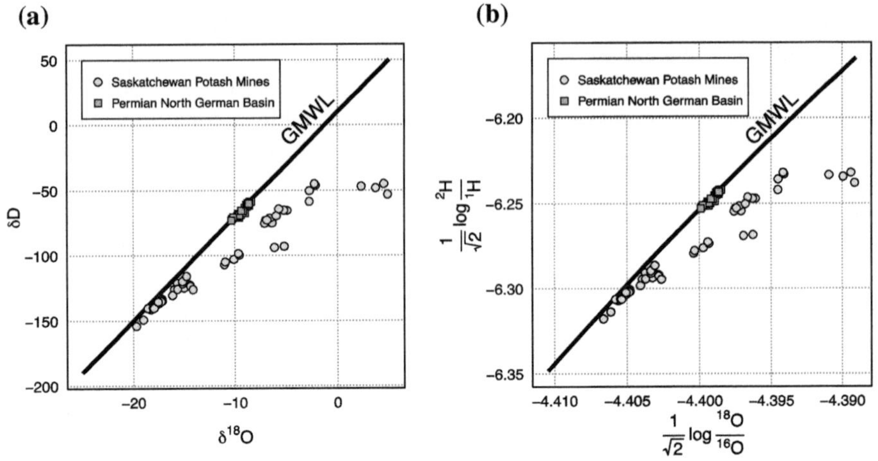

Fig. 1 H and O isotopes of water for North German basin brines [18], presented **a** in standard delta notation and **b** as a four-part ilr transformation. The *solid black line* labelled GMWL is the global meteoric water line, which represents the isotopic composition of terrestrial surface and near surface waters, and is dominantly a function of latitude, elevation, and temperature. Waters that have experienced evaporation or re-equilibration with minerals in the subsurface tend to plot to the right of the GMWL. Note the slight curvature of the GMWL in Fig. 1b

show secondary processes such as albitization or incorporation of clays. Engle and Blondes [6] used ilr transformations of ions in produced waters from the Permian Basin, USA, to link geochemical variation to thermodynamic mineral equilibration models. The incorporation of isotopes into such CoDA-based studies would be an important advancement for the study of basinal brine geochemistry.

Figure 1a shows hydrogen and oxygen isotopes in water for brines from the North German Basin [18] and from potash mine shafts in Canada [17]. The global meteoric water line (GMWL) represents the isotopic composition of terrestrial surface and near surface waters [5]. The position of a given data point along that line is a function of many variables, dominantly latitude, elevation, and temperature. Evaporative loss, which preferentially removes the lighter isotopes (^{1}H and ^{16}O), results in "heavier" δ^{18}O and δ^{2}H values that plot to the right of the GMWL. Exchange of oxygen between water samples and typically isotopically heavy minerals such as carbonate and clay minerals, can also produce enrichment of ^{18}O, causing affected samples to plot to the right of the GWML. Deep formation brines that previously experienced evaporation or isotopic exchange with carbonate and silicate minerals but have not interacted with surface waters tend to plot in this region.

The Permian North German Basin brines [18] plot directly on the GMWL, suggesting they are of meteoric origin. The formation waters from the potash mine shafts in Saskatchewan [17] have a wide range in isotopic composition that is nearly as broad as all known terrestrial water compositions [21]. This makes it particularly useful to detect differences between the traditional representation and a similar one created using the ilr transform of a four-part subcomposition, with each part being the individual isotopic concentrations. At first glance, the plots look identical except for

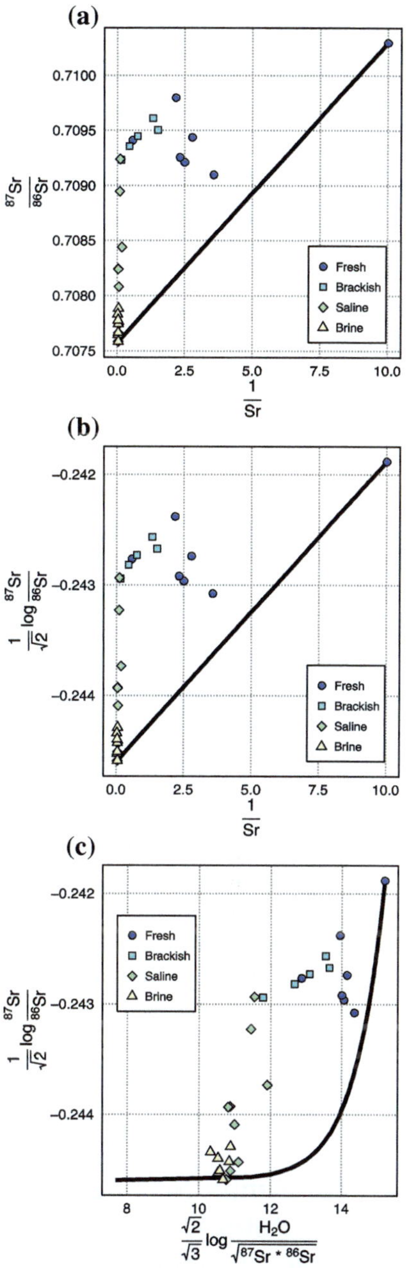

Fig. 2 Sr isotope plots of North German Basin brines [18]. The *solid black line* represents linear mixing between the most fresh and the most saline samples. **a** Traditional presentation of strontium isotopes in which mixing is linear. **b** Two-part ilr transformation for the y-axis using ^{87}Sr and ^{86}Sr. **c** Three-part ilr transformation using ^{87}Sr, ^{86}Sr, and H_2O

the values on the axes, suggesting that the difference is simply one of scale. However, it is apparent from a closer look at Fig. 1b that the GMWL shows some curvature at the top right of the plot. Though these differences do exist, given that these data span nearly the entire range in observed values for the isotopic composition of natural waters, it is small enough to be negligible and the ilr transformation does not appear to offer any major practical benefits in this type of plot for exploratory analysis.

We can also examine radiogenic $^{87}Sr/^{86}Sr$ isotopes in a similar way. Figure 2a again shows the Kloppman et al. [18] data set, color-coded by salinity where fresh is <1 g/L TDS, brackish is 1–10 g/L TDS, saline is 10–100 g/L TDS, and brine is >100 g/L TDS. This plot of $^{87}Sr/^{86}Sr$ versus 1/Sr is a standard way to show mixing between Sr isotopic compositions, as pure mixtures of two end-members will fall along straight lines. The black line in this figure is a hypothetical linear mixture between the most fresh and the most saline (brine) sample in the data set. One can quickly infer that these samples do not derive their chemical and isotopic variation from simple mixing of these two end-members, yet because the data have not been transformed to a Euclidean geometry, the distances from the data to the model may be distorted. Using a two-part ilr transformation for the Sr isotopes ^{87}Sr and ^{86}Sr (Fig. 2b), the same scaling relationship observed in Fig. 1b is apparent, whereby the log ratio of a small variation is simply that small variation scaled. Figure 2c shows a three-part ilr transformation with ^{87}Sr, ^{86}Sr, and H_2O. Similar interpretations of mixing can be drawn by comparing the data to the models in real space, yet the mixing relationship is no longer linear. While not as simple to the naked eye, the fit of the mixing model is only truly quantifiable in Fig. 2c because distances are preserved through the ilr transformation.

The result that the ilr-transform of subcompositions with small log ratio variance is simply a scaling (or a perturbation) of the ratio has important implications for the interpretation of both isotopes and trace elements. Unlike major element ratios (e.g. Ca/SO_4), which are more likely to have large log ratio variance and are subject to significant spurious correlations, trace element and isotope ratios may not be subject to the same constraints. In the following section we attempt to quantify under what conditions the ilr transformation of compositional ratios functions as a simple perturbation. For a simple perturbation, exploratory analysis of trends will be identical whether simple ratios or the ilr transformation is used, although distances will be different. We further address how major versus trace elements fit into this context, i.e. at what point does a trace element become trace enough, relative to another element, to not be subject to the constraints that require a log ratio transformation for proper interpretation.

4 The Simple Scaling Effect of Ilr Transformations on Low Log Ratio Variance (Isotope) or Low Concentration (Trace) Components

Figure 3 schematically examines the effect of a two-part ilr transformation on any pair of elements, x_1 and x_2. The x-axis represents a ratio of any two parts before transformation, including isotopic ratios (e.g. $^2H/^1H$ or $^{87}Sr/^{86}Sr$) or elemental ratios

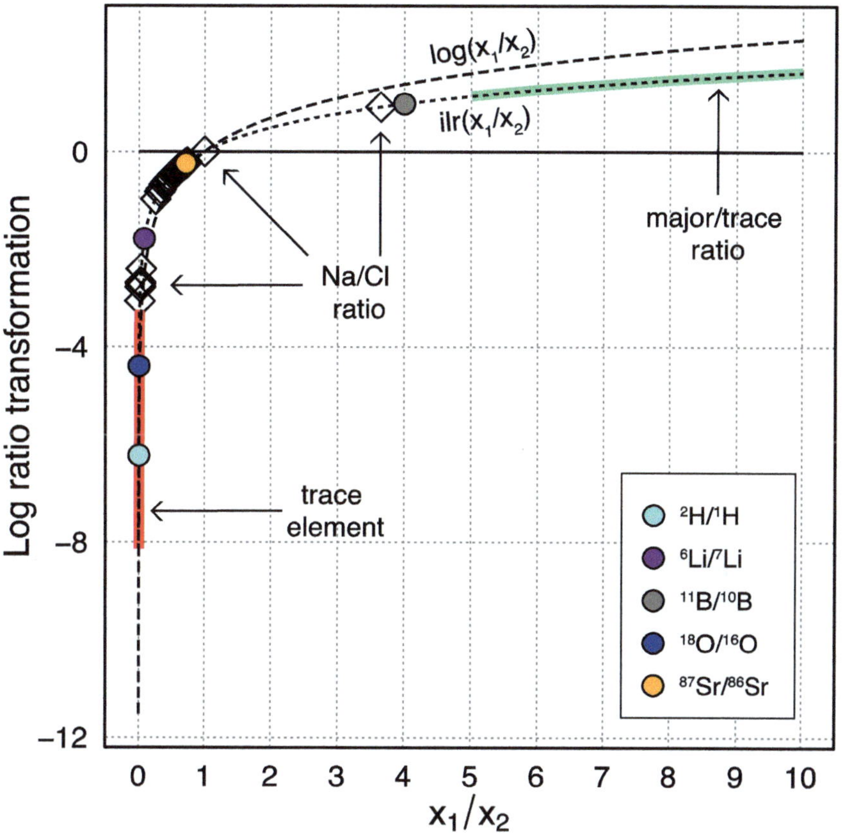

Fig. 3 Two-part log ratio transformations of two parts (x_1 and x_2) of a composition. The *dashed line* is the natural log of this ratio, and the *dotted line* is the ilr transformation of this ratio ($z_{two\text{-}part}$) defined by Eq. 3. The *colored circles* represent the full extent of variation for a given isotope ratio. The *red bar* is a schematic representation of the ilr variability for trace elements (x_1 = [trace element], x_2 = [solution]) or a ratio of a trace element to a major component. The *green bar* is a schematic representation of the ratio of a major component to a trace component. The black diamonds are the Na/Cl ratios from [17] that represent the variation of two components of similar concentration

(e.g. Cl/Br or Na/Cl). It can also represent simple trace element concentrations (e.g. Sr/solution or B/solution) because they approximate a trace/major ratio (e.g. Sr/H_2O or Sr/Cl). This means that we can consider the simple trace element concentration to be equivalent to a ratio of a trace to a major component of a system. The y-axis represents the ilr transformation of that ratio (Eq. 3):

$$z_{two\text{-}part} = \frac{1}{\sqrt{2}} \log \frac{x_1}{x_2}. \tag{3}$$

The dashed curve is a log ratio transformation, and the dotted line is a two-part ilr transformation, the only difference being the coefficient in the ilr transformation. The colored circles represent different isotopic systems, centered at their standard value (e.g. VSMOW) plotted on the ilr curve. The typical variation of an isotope ratio is small relative to the scale of the plot and falls within the bounds of the circle. For example, the typical $^{87}Sr/^{86}Sr$ ratio for natural waters is 0.70–0.72, which is just a small region within the orange circle. The most important interpretation from this plot is that wherever some subset of data is linear at a given scale along the log ratio or ilr curves, the log ratio transformation of the data simply corresponds to a perturbation or scale change of the original ratio. For the isotopic systems, the variation within the ratio is so small relative to the scale of the plot that it functions as a tangent line to the curve. This explains why there is little visible difference between Fig. 1a with Figs. 1b and 2a with Fig. 2b. This low log ratio variance corresponds to the high "stability" of Filzmoser et al. [11], a compositional measure of proportionality between two components.

Another observation from Fig. 3 is that small log ratio variance is not the only way an ilr transformation reduces to a simple perturbation. The log ratio and ilr curves function as linear far from $x_1/x_2 = 1$ (as x_1/x_2 approaches either 0 or $+\infty$), even if the ratios have high log ratio variance. For example, the green segment could represent Cl/Br ratios from 5/1 to 10/1. A similar green segment would plot for Cl/Br ratios of 50/1–100/1 or 500/1–1000/1 (not shown). This relationship would hold for many produced water samples given that the concentration of Cl is typically orders of magnitude greater than Br. Thus the log ratio of such Cl/Br data may function as a simple perturbation or a scaling of the axis. The same result occurs for a component pair with orders of magnitude log ratio variance but ratios much lower than $x = 1$, shown by the red segment. The red segment could represent the relationship between a trace and a major element or just a trace element concentration, using the approximation for trace elements described above. Any high log ratio variance pair would be considered to have low stability by Filzmoser et al. [11] but here we show that there are certain conditions common in geochemical data (trace element concentrations and major element/minor element ratios over a restricted range) where a low stability ratio will function as a high stability ratio through an ilr transformation.

Any ratios made from parts that have similar concentrations, including most major components, will plot around $x_1/x_2 = 1$ where the curvature is most apparent. For example, Na/Cl ratios of the Saskatchewan potash mine waters (open black diamonds in Fig. 3) fall along the most curved part of the trend, where an ilr transformation will have the greatest effect on the shape of the data. This is why such large differences are seen in major element ternary diagrams when log transformations are applied. Though these types of diagrams are often used for classification and a general understanding of the major components in a system, most robust interpretations of process are made by combining isotopic ratios (colored circles) and trace element concentrations (which again approximate the ratios of trace to major components and are represented by the red segment), both of which fit the conditions for an ilr transformation providing less obvious extra benefit for exploratory trend analysis.

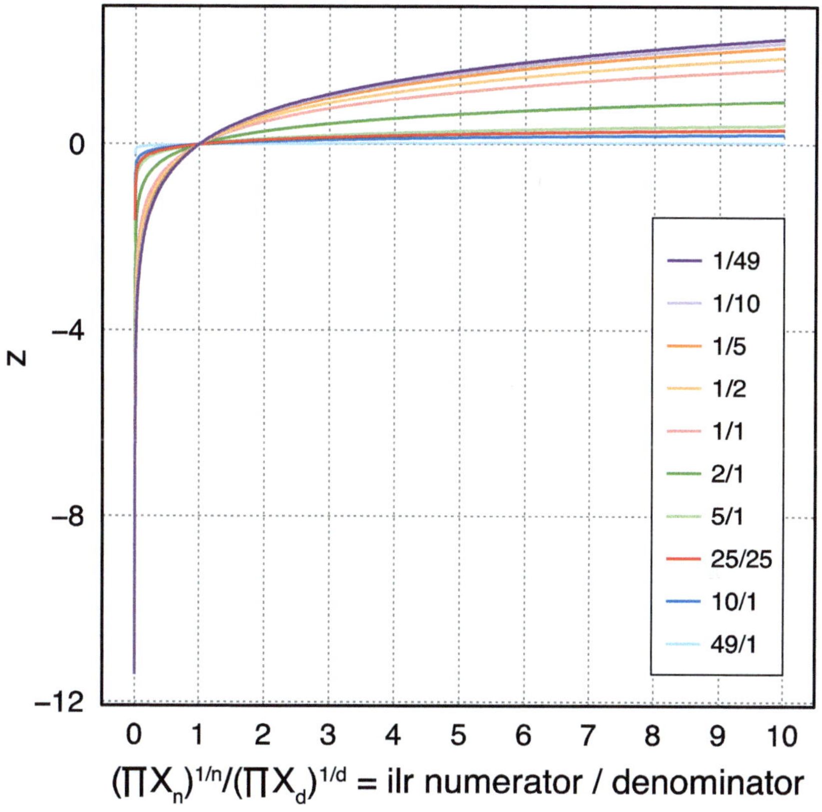

Fig. 4 Multivariate ilr transformation for a 50-part composition for a range of binary partitions. The x-axis is the ratio of the geometric mean of the numerator and denominator in the ilr equation

This generalization about the relative impact from conversion of ratios to an ilr basis can be further expanded to a multivariate ilr approach. Figure 4 is similar to Fig. 3, but instead of a ratio of two elements, the x-axis now represents the ratio of the geometric mean of the +1 balance components to the −1 balance components in the ilr equation (Eq. 4):

$$z = \frac{\sqrt{n \times d}}{\sqrt{n + d}} \log \frac{(\prod X_n)^{\frac{1}{n}}}{(\prod X_d)^{\frac{1}{d}}}, \quad (4)$$

where X_n and n are the components and the number of components, respectively in the +1 balance or the numerator, and X_d and d are the components and the number of components in the −1 balance or the denominator, respectively.

The numerator is the geometric mean of all of the positive balance parts and the denominator is the geometric mean of all the negative balance parts. The geometric mean of the concentrations of the (+) balances equals the geometric mean of the concentrations of the (−) balances at $x = 1$. The different curves show the effect of changing the grouping of the balances. The x-axis represents concentration ratios and the different curves represent the ratio of the number of parts. Typical geochemical data sets are large, and a 50-part data set would have $n(n-1)/2 = 1225$ possible binary partitions. The different curves in Fig. 4 show the range of binary partition possibilities including one part against the rest (1/49 = one part in the numerator and 49 parts in the denominator or 49/1 = 49 parts in the numerator and one part in the denominator), evenly split parts (25/25 = 25 parts in the numerator and 25 parts in the denominator), and smaller subcompositions (5/1, 1/5, 1/1, etc.). The 1/1 curve in Fig. 4 is the same as the ilr curve in Fig. 3 for a two-part subcomposition. As in Fig. 3, the curvature of these lines can be used as a proxy for whether an ilr transformation will change the shape of any data trends beyond a simple scale-change. If the spread of data for an isotopic or chemical system falls into a linear region of the curve, the ratio of the geometric means and ilr plots will be nearly the same. Putting nearly all parts into the numerator partition creates a very rapid slope change from nearly vertical at low X to nearly flat just above z=0. Putting nearly all parts into the denominator creates a slow changing slope that shows the most curvature of the range of potential geochemical concentrations. The more parts that are used, the less curvature that exists, even for evenly split binary partitions (e.g. compare the 25/25 to the 1/1 curve). For a geochemical data set, a first partition might be splitting the cations and anions. In this example, the curve is almost completely linear except for the rapid change at $X = 1$, meaning that unless X is close to unity, the first sequential binary partition would generate an ilr transformation that is a simple perturbation of the ratio of the geometric means of each part. This is less analogous to traditional ratio plotting methods than the two-part ilr described above since it is not typical to plot the ratio of geometric means, but it still helps underscore stability in multivariate systems.

5 Clr-Biplot Interpretation of Low Log Ratio Variance (Isotope) or Low Concentration (Trace) Components

Isotopic data can also be incorporated into biplots, but unlike trace elements with large log ratio variance, isotopes with low log ratio variance cannot be easily discerned from one another. Principal component analysis of clr-biplots are useful for showing multivariate compositional data on a single plot [2]. When derived from the covariance matrix of clr-transformed variables, the length of the links between the ray end points approximate the relative log ratio variance between those two variables. Using data from Kloppman et al. [18], Fig. 5a shows covariance clr-biplot of only TDS and all of the O and H isotopes of water. The H_2O concentration is also shown for illustrative purposes to compare the individual isotope rays to the H_2O ray, but it is a redundant variable because $[H_2O] = [^{18}O] + [^{17}O] + [^{16}O] + [^{2}H] + [^{1}H]$.

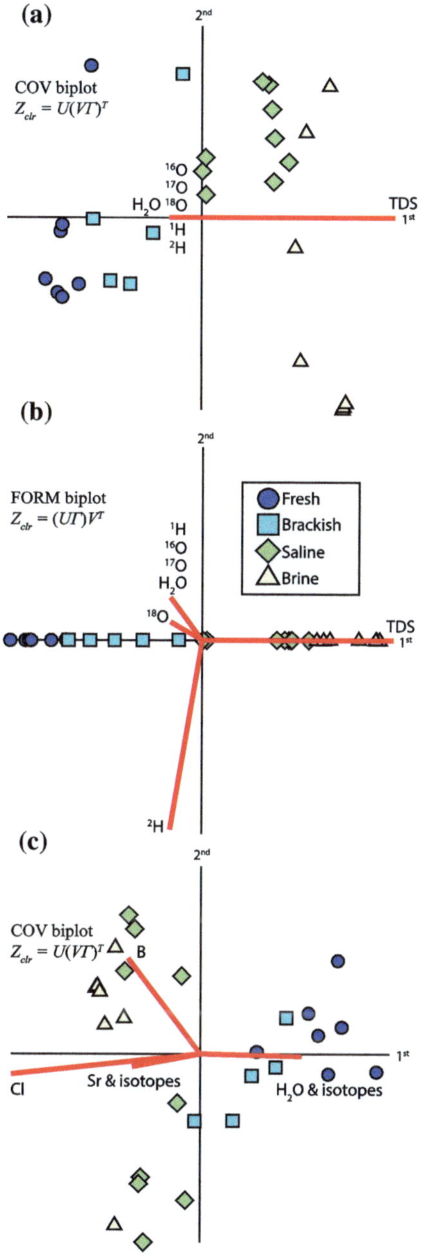

Fig. 5 Clr-biplots of Kloppman et al. [18] data. Only TDS, H_2O and the O and H isotopes are plotted. **a** Covariance biplot. **b** Form biplot. **c** Covariance biplot with major component Cl and trace component B added

All isotopes of O and H in water lie on the same ray as H_2O, which is unsurprising as their concentrations are completely dependent on one another and sum to H_2O. The scores are more clearly seen in the "form" version of the biplot (Fig. 5b), in which the relationship between the scores and the coordinate axes are more visible. Here we see that the 1st coordinate axis contains the entirety of the variability and represents a salinity trend. Though the rays and the links between them have less meaning in the form biplot, small deviations are visible between the numerators in the isotopic ratios (^{18}O, ^{2}H) and the ray that represents H_2O, ^{16}O, ^{17}O, and ^{1}H. The longer ray between ^{1}H and ^{2}H relative to the ray between ^{18}O and ^{16}O on the biplot supports the theory and observation that the larger mass differences between isotopes in the hydrogen system (^{2}H has twice the mass of ^{1}H) produces more isotopic fractionation than the oxygen system (^{18}O only has 12.5 % more mass than ^{16}O). Nearly 100 % of the variance is found in the link between TDS and H_2O and the scores project along this link as one would expect according to salinity. Figure 5c shows a subcomposition including trace (Sr, B) and major (H_2O, Cl) components. The H_2O–Cl link is similar to the H_2O–TDS link in Fig. 5a in that it represents variation related to salinity. The scores project onto this link as a function of salinity. The links between isotopic pairs are again near zero. The different isotopes of O and H plot on the H_2O concentration ray, and similarly the different isotopes of Sr plot on the Sr concentration ray. Unlike the isotopic ratios, trace elements have high log ratio variance with other trace elements (Sr–B) and with major elements (Cl–B). In this case, what is controlling the log ratio variance between Sr and B is not related to the salinity. Clr-biplots are useful for determining trends and groups within homoscedastic data, but for isotopes with low log ratio variance, scaling parts to similar variance is necessary for interpretation (e.g. Puig et al. [20]).

6 Discussion and Conclusions

Much of the work on compositional data analysis in the geosciences has focused on major element variation, whether mineral stoichiometry (e.g. Grunsky et al. [14]), the major ions in an aqueous solution (e.g. Buccianti and Pawlowsky-Glahn [3]), or whole rock geochemistry (e.g. Geboy et al. [12]). Incorporating trace elements properly into CoDA has been increasingly used with success, particularly for spatial applications (e.g. Grunsky et al. [15]; Tolosana-Delgado and van den Boorgaart [25]). It has been shown that stable isotopes can be included by scaling the ratios [20, 24]. We show here that not only can radiogenic isotopes be included in a similar way but that we can more generally address non-major parts of a system in two ways: either as (1) low log ratio variance parts (isotopic ratios) or (2) low concentration parts (trace elements).

Low log ratio variance parts, like an isotopic ratio, will be all but invisible on a clr-biplot without prior scaling (Fig. 5a). The ilr transformation of isotopic ratios produces interpretable trends, but at the scale of analysis these trends are nearly identical to traditional presentations of isotopic ratios (Figs. 1 and 2). This is because the log ratio of two low variance parts is simply a linear scaling, or a perturbation, of that ratio (Fig. 3). Any ratio with low variance will function as a tangent line to the log curve, meaning that the interpretive power of the ilr transformation is similar to that of the simple log ratio (Fig. 3). This can be extrapolated to multivariate full- or sub-compositions that represent a large geochemical data set. For a data set with 50 parts (all major and trace elements) split evenly between, for example, cations and anions, the ilr transformation will be a simple perturbation of the ratio of the geometric means except near where the geometric mean of the cations in the numerator is vanishingly small relative to the denominator (Fig. 4). In practice, this should never happen for a properly charge balanced analysis, but it might occur if the $+1$ balance represented the 25 lowest concentration trace elements and the -1 balance represented the 25 highest concentration components. Though low log ratio variance parts like isotopes will likely always be a simple perturbation in ilr space, there are other geologic data scenarios where this can be the case as well. One clear example is low concentration parts, such as trace elements. The vertical red bar in Fig. 3 represents an ilr transformation of a trace element. At the scale of analysis, particularly when compared to the Na/Cl ratio, the ilr transformation of a trace element is a simple perturbation. This is also the case for large positive ratios over a restricted range (Fig. 3, green bar).

When geoscientists choose to make interpretations based on geochemical variation, it is important to avoid spurious correlations and inappropriate statistical models by properly transforming the data. One could claim that an orthonormal log ratio transformation covers all statistical bases and should be used for every application. In fact, the ilr transformation has been shown to provide powerful insight to the interpretation of brine geochemistry where traditional methods could not, including identifying spurious mixing trends [8], mineral equilibrium relationships [6] and the effects of diffusion into clays and input of ions from kerogen maturation [7]. In practice, however, there are many cases where the ilr transformation results in nearly the same data patterns as the traditional approach, in which the axes are much more familiar (such as specific values for isotopic ratios). The analysis here shows that traditional approaches will generally yield the same results as an ilr transformation for isotope ratios, trace elements compared to major elements or the entire solution, and certain special cases of major element ratios.

Acknowledgments Funding for this project was provided by the U.S. Geological Survey Energy Resources Program. The authors appreciate constructive reviews by Ricardo Olea and two anonymous reviewers, as well as useful feedback from the CoDaWork 2015 workshop participants and Allan Kolker.

References

1. Aitchison, J.: The Statistical Analysis of Compositional Data (Reprinted in 2003 by The Blackburn Press), p. 416. Chapman & Hall Ltd., London (UK) (1986)
2. Aitchison, J., Greenacre, M.: Biplots of compositional data. Appl. Stat. **51**, 375–392 (2002)

3. Buccianti, A., Pawlowsky-Glahn, V.: New perspectives on water chemistry and compositional data analysis. Math. Geol. **37**, 703–727 (2005)
4. Chayes, F.: On correlation between variables of constant sum. J. Geophys. Res. **65**(12), 4185–4193 (1960)
5. Craig, H.: Isotopic variations in meteoric waters. Science **133**, 1702–1703 (1961)
6. Engle, M.A., Blondes, M.S.: Linking compositional data analysis with thermodynamic geochemical modeling: oilfield brines from the Permian Basin, USA. J. Geochem. Explor. **141**, 61–70 (2014)
7. Engle, M.A., Reyes, F.R., Varonka, M.S., Orem, W.H., Ma, L., Ianno, A.J., Schell, T.M., Xu, P., Carroll, K.C.: Geochemistry of formation waters from the Wolfcamp and "Cline" shales: insights into brine origin, reservoir connectivity, and fluid flow in the Permian Basin, USA. Chem. Geol. **425**, 76–92 (2016)
8. Engle, M.A., Rowan, E.L.: Interpretation of Na–Cl–Br systematics in sedimentary basin brines: comparison of concentration, element ratio, and isometric log-ratio approaches. Math. Geosci. **45**(1), 87–101 (2013)
9. Egozcue, J., Pawlowsky-Glahn, V., Mateu-Figueras, G., Barceló-Vidal, C.: Isometric log ratio transformations for compositional data analysis. Math. Geol. **35**, 279–300 (2003)
10. Faure, G., Mensing, T.M.: Isotopes: Principles and Applications, 3rd edn, p. 928. John Wiley & Sons, Inc. (2014)
11. Filzmoser, P., Hron, K., Reimann, C.: The bivariate statistical analysis of environmental (compositional) data. Sci. Tot. Environ. **408**(19), 4230–4238 (2010)
12. Geboy, N.J., Engle, M.A., Hower, J.C.: Whole-coal versus ash basis in coal geochemistry: a mathematical approach to consistent interpretations. Int. J. Coal Geol. **113**, 41–49 (2013)
13. Goldschmidt, V.M.: Geochemistry, Muir, A. (ed.) Clarendon Press, Oxford (1954)
14. Grunsky, E.C., Kjarsgaard, B.A., Egozcue, J.J., Pawlowsky-Glahn, V., Thió i Fernández de Henestrosa, S.: Studies in stoichiometry with compositional data. In: 3rd compositional data analysis workshop CoDaWork '08, p. 7 (2008)
15. Grunsky, E.C., Mueller, U.A., Corrigan, D.: A study of the lake sediment geochemistry of the Melville Peninsula using multivariate methods: applications for predictive geological mapping. J. Geochem. Explor. **141**, 15–41 (2014)
16. Irving, A.J.: A review of experimental studies of crystal/liquid trace element partitioning. Geochimica et Cosmochimica Acta **42**(6), 743–770 (1978)
17. Jensen, G.K.S., Rostron, B.J., Duke, M.J.M., Holmden, C.: Chemical profiles of formation waters from potash mine shafts, Saskatchewan. In: Summary of Investigations 2006, Volume 1, Saskatchewan Geological Survey, Sask. Industry Resources, Misc. Rep. 2006-4.1, CD-ROM, Paper A-7, p. 8 (2006)
18. Kloppman, W., Négrel, P., Casanova, J., Klinge, H., Schelkes, K., Guerrot, C.: Halite dissolution derived brines in the vicinity of a Permian salt dome (N German Basin). Evidence from boron, strontium, oxygen, and hydrogen isotopes. Geochimica et Cosmochimica Acta **65**, 4087–4101 (2001)
19. Pearson, K.: Mathematical contributions to the theory of evolution. On a form of spurious correlations which may arise when indices are used in the measurements of organs. Proc. R. Soc. Lond. **60**, 489–498 (1897)
20. Puig, R., Tolosana-Delgado, R., Otero, N., Folch, A.: Combining isotopic and compositional data: a discrimination of regions prone to nitrate pollution. In: Pawlowsky-Glahn, V., Buccianti A. (eds.) Compositional Data Analysis: Theory and Applications, pp. 302–317. John Wiley & Sons, Ltd (2011)
21. Rozanski, K., Araguás-Araguás, L., Gonfiantini, R.: Isotopic patterns in modern global precipitation. In: Swart, P.K., Lohmann, K.C., McKenzie, J., Savin, S. (eds.) Climate Change in Continental Isotopic Records, pp. 1–36. American Geophysical Union, Washington, D.C. (1993)
22. Sofer, Z., Gat, J.R.: Activities and concentrations of oxygen-18 in concentrated aqueous salt solutions: analytical and geophysical implications. Earth Planet. Sci. Lett. **15**, 232–238 (1972)

23. Sofer, Z., Gat, J.R.: The isotope composition of evaporating brines: effect of the isotopic activity ratio in saline solutions. Earth Planet. Sci. Lett. **26**, 179–186 (1975)
24. Tolosana-Delgado, R., Otero, N., Soler, A.: A compositional approach to stable isotope data analysis. In: 2nd Compositional Data Analysis Workshop CoDaWork '05, p. 11 (2005)
25. Tolosana-Delgado, R., van den Boogaart, K.G.: Towards compositional geochemical potential mapping. J. Geochem. Explor. **141**, 42–51 (2014)

Towards the Concept of Background/baseline Compositions: A Practicable Path?

A. Buccianti, B. Nisi and B. Raco

Abstract Water geochemistry is often investigated considering a large number of variables, including major, minor and trace elements. Some of these are usually well associated due to coherent geochemical behaviour, but the effect of anthropic factors tends to increase data variability, sometimes obscuring the natural laws governing their relationships. It may thus be difficult to identify geochemical features linked to natural phenomena, as well as to separate geogenic anomalies from the anthropogenic ones, or to define background or baseline concentrations for single chemical elements. This is particularly true at regional level, where numerous phenomena may interact and mix together, forming a complex pattern not easy to interpret. The identification of background or baseline values is particularly difficult due to the compositional nature of chemical variables, so that under the Compositional Data Analysis (CoDA) theory single background or baseline values lose their meaning. However, they are fundamental references for public institutions and government policies. In this contribution a new approach is proposed, aimed at investigating the regionalised structure of the geochemical data by considering the joint behaviour of several chemical elements. The approach is based on the robust CoDA theory, so that the proportionality features of abundance data are fully taken into account, enhancing their relative multivariate behaviour, as well as the influence of outliers. An application example is presented for the groundwater compositions in Tuscany Region, a surface of about 23,000 km^2, where more than 6000 wells have been sampled and analysed. The mapping of robust Mahalanobis distance was able to indicate (1) in which part of the investigated area the pressure toward anomalous behaviour was higher, (2) where the compositions nearest to the barycentre were and (3) if spatial continuity was present in limited portions of the territory.

A. Buccianti (✉)
Department of Earth Sciences, University of Florence, Via G. La Pira 4,
50121 Firenze, Italy
e-mail: antonella.buccianti@unifi.it

B. Nisi · B. Raco
CNR-IGG (Institute of Geosciences and Earth Resources), via G. Moruzzi, 1, 56124 Pisa, Italy

J.A. Martín-Fernández and S. Thió-Henestrosa (eds.), *Compositional Data Analysis*, Springer Proceedings in Mathematics & Statistics 187,
DOI 10.1007/978-3-319-44811-4_3

Keywords Water chemistry · Compositional data analysis · Baseline · Fractals · Dissipative structures

1 Introduction

1.1 Background or Baseline: A Summary

Geogenic or "natural background" substances in the environment are known to occur as concentrations in air, soils and waters. A number of terms are used to convey the expected concentrations of an element in natural materials. These include: normal, typical, baseline, ambient, characteristic, natural, background and widespread. There are some subtle differences between these terms in literature, they can mean different things in different disciplines, and they can be confused with alternative uses [29].

In environmental geochemistry, background is a relative measure to distinguish between natural element or compound concentrations and anthropogenically influenced concentrations in real sample collectives. However, as reported in Nordstrom [30] it may be difficult for water chemistry to refer to natural background as unpolluted or pristine preindustrial conditions. This happens because (1) a widespread global contamination by several trace constituents has occurred, (2) natural variations can be large, so that a single analytical result for a given element or compound cannot be useful and (3) the effect of scale and study objectives could have played a very important role in the determination of background [33]. One of the most interesting discussions on this subject is reported in Matschullat et al. [27]. These authors recognise that the citation of single values for a geochemical background is neither useful for the identification of the geogenic contribution nor for the determination of an anthropogenic contamination, because single values do not yield information about natural deviation.

Another term, geochemical baseline, is often closely associated with geochemical background, terms often used as synonyms (e.g. [20, 28]). However, the geochemical baseline is often considered as the natural background in diffusely polluted areas where the latest term cannot be further defined because natural conditions are compromised.

1.2 Baseline Hydrochemical (Compositional) Facies

The knowledge of groundwater chemistry and of associated background/baseline values for elements and compounds is a priority for human health as established by the environmental organisms and institutions of many industrialised and developing countries [35, 38]. Notwithstanding the importance of this item, the definition of the background or baseline content in water is difficult and cannot be based only on a good analytical phase and on a time-limited sampling campaign. In fact, water chemistry can vary in space depending on changing climate conditions as well as on the contribution of anthropogenic pollution. Moreover, in the same place the abundance of chemical species can change in time due to the effect of several environmental and anthropic factors [30].

Consequently, it is evident that single concentration values are not able to give representative information about the influence of complex phenomena as occurring in groundwater systems. In this natural reservoir, only simultaneous chemical equations are able to describe the different mineral/water equilibria and the investigation of data variability assumes a fundamental role. For these reasons, it is our opinion that only a compositional approach is able to describe the conditions of such a complex phase [3, 7, 9]. Moreover, due to the expected presence of anomalous values, robust methods can improve our proposal [11, 26, 34, 36].

Thus, the combination of compositional and robust methods applied to water chemistry in anthropic areas could force the concept of baseline for single values to evolve to that of baseline composition or baseline hydrochemical (compositional) facies. The approach can be extended to different scales in groundwater investigation and represents an implementation, moving from the value of single variables to that of compositions when the joint distribution of D variables is considered [8, 12].

2 Materials and Methods

2.1 Data Sources

The geochemical data used in this study refer to groundwater samples actually stored in the GEOBASI database[1] [32] (http://www506.regione.toscana.it/geobasi/index.html) representing the official repository for the geochemical composition of geological media of the Tuscany Region (Central Italy), a surface of about 23,000 km^2. The northern and southern sectors of Tuscany are bordered by the Northern Apennine, an orogenic belt formed by the Cretaceous to Miocene compressive phase related to the collision of the European and African plates. The metamorphic Paleozoic basement (e.g. phyllitic to quarzitic and micaschists rocks to Triassic evaporitic anyhydrites), the Mesozoic and Cenozoic carbonate and evaporitic formations, overlain by flysch series, as well as granite intrusions and volcanic rocks are typical lithologies [10].

The chemical compositions of 6,808 cases (springs and wells) were considered, 6,435 of which were geo-referenced and checked through the inspection of the original field maps. For those samples where the main composition (Ca, Mg, Na, K, HCO_3, SO_4 and Cl) was available, the quality of the geochemical data was checked by means of a simple charge balance and only those waters having a percentage deviation <10 (4,804 samples) were taken into account for further processing.

[1]Part of Consorzio LaMMA, http://www.lamma.rete.toscana.it.

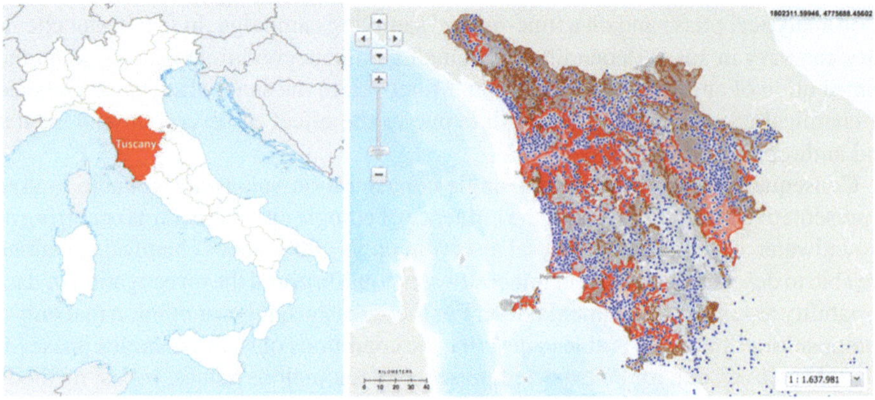

Fig. 1 Location of all the water samples for which the chemical composition is stored in the GEOBASI repository [32]. Cases located in *red* and *brown* areas correspond to groundwater located in alluvial and fracture rock aquifers, respectively

For the sake of clarity, it should be noted that all the geochemical data were used, including those repeatedly analysed over time, so that variability could be also affected by seasonality (Fig. 1). Besides this source of variability, other phenomena such as seawater intrusion, land use, pollution, are expected to affect data variability even if with a more punctual spatial impact.

2.2 *Statistical Methodologies*

In this contribution, our aim is to shift the attention from the investigation of single variables of compositional data sets to their joint multivariate behaviour. To achieve this target, the principles of CoDA (compositional data analysis) theory were followed [4], combined with the application of multivariate robust methodologies [11, 26]. Compositional data are vectors of positive values quantitatively describing the contribution of D parts of some whole, which carry only relative information [3, 4]. Due to these features, the approach based on the Euclidean geometry of the real space to the statistical analysis of compositions may give misleading results, since compositional data pertain to the simplex sample space [6, 7, 13]. The simplex sample space is governed by the Aitchison geometry, and has all the properties of a $(D-1)$ dimensional Euclidean space [13]. To work in these unconstrained conditions, compositions need to be expressed as vectors of values that belong to such a space.

In our case the isometric log-ratio (ilr) transformation, proposed by Egozcue et al. [15], was adopted. Notwithstanding its theoretical advantages and practical properties, its use may be compromised when coordinates have to be interpreted from a geochemical point of view. However, if the concept of balance between groups of parts originated by a sequential binary partition is considered [14], this path may be highly simplified. In a sequential binary partition, in each of the $D-1$ steps of the procedure, the compositional parts are divided into two non-overlapping groups; the resulting $D-1$ ilr-variables represent balances between these groups in R^{D-1}:

$$ilr_i = \sqrt{\frac{r \times s}{r+s}} \log \frac{g(c_+)}{g(c_-)}, \tag{1}$$

with $i = 1, 2, \ldots, D-1$ and where $g(c_+)$ represents the geometric mean of the r variables of the numerator of the balance, $g(c_-)$ the geometric mean of the s variables of the denominator.

The matrix of ilr coordinates was analysed by robust methods with the aim to identify samples with anomalous behaviour but simultaneously avoiding their effect on the classical estimates. For this purpose, the Mahalanobis robust distance of a composition, labelled RD_i, was used to detect whether it was an outlier composition or not. It was defined as:

$$RD_i = \sqrt{(x_i - \widehat{\mu}_{MCD})^T\, \widehat{\Sigma}_{MCD}^{-1}\, (x_i - \widehat{\mu}_{MCD})}, \tag{2}$$

with $\widehat{\mu}_{MCD}$ and $\widehat{\Sigma}_{MCD}$ the location and scatter estimates obtained by using the Minimum Covariance Determinant (MCD) estimator [34, 36]. Mahalanobis distance identifies observations that lie far away from the centre of the data cloud, giving less weight to variables with large variances or to groups of highly correlated variables. This distance is often preferred to the Euclidean distance which ignores the covariance structure and treats all variables equally.

The robust distance is an improvement on the Mahalanobis one, where classical mean and empirical covariance matrix are used as estimates of location and scatter [16–19]. Under the normal assumption the outlier compositions are those compositions having a robust distance larger than the chosen cut-off value, here taken as $\sqrt{\chi^2_{D-1,0.975}}$.

This approach, however, does not account for the sample size n of the data, and independently from the data structure observations could be flagged as outliers even if they belong to the data distribution. A better procedure could use a fixed threshold to be adjusted to the data set at hand. The chi-square plot is often used for this purpose by plotting the squared robust Mahalanobis distances against the quantiles of χ^2_D, then by deleting the most extreme points, identified as outliers, until the remaining points follow a straight line [21].

This study starts from the work of Filzmoser and Hron [17] and proposes some original improvements considering the investigation of the distributional form of RD_i and its spatial behaviour. All the analyses have been performed using robust routines developed in Matlab and R [31, 36].

The mapping of robust Mahalanobis distance could be able to indicate (1) in which part of the investigated area the pressure toward anomalous behaviour is higher, (2) where the compositions nearest to the barycentre are and (3) if spatial continuity is present in limited portions of the territory. Important information about the geochemical processes, their diffusion and influence in the different parts of the investigated areas can be consequently obtained. To be stressed here is the fact that the investigated area, corresponding to a surface of about 23,000 km^2, presents a sample point pattern that cannot be described by a Complete Spatial Randomness (CSR) model (goodness-of-fit testing for CSR with $p < 0.05$, spatstat R-package, [31]). The wells were drilled where needed and where geological conditions guaranteed a successful production.

3 Results and Discussion

3.1 Sequential Binary Partition of Water Chemistry

In the investigation of the main groundwater composition of Tuscany region, the first step was to transform the original variables expressed in mg/L by using the isometric log-ratio conversion. To achieve this aim, the sequential binary partition proposed by Egozcue and Pawlowsky-Glahn [14] was adopted. In each of the $D-1$ steps of the procedure the compositional parts were split into two non-overlapping groups; the resulting $D-1$ ilr-variables represented balances between these groups in R^{D-1}. Thus, the balances between two groups of parts are orthogonal log-ratio contrasts between geometric means of the selected non-overlapping groups. The sequential binary partition of [Eq. (1)] was chosen so that all the cations were balanced (symbol |) considering the following steps:

[Ca, Mg, K, Na | HCO, Cl, SO_4],
[Ca, Mg | K, Na],
[Ca | Mg],
[K | Na],
[HCO_3 | Cl, SO_4],
[Cl | SO_4].

In this conversion, the geometric means are central values in each group of parts, their ratio measures the relative weight of each group and the logarithm provides the appropriate scale; the square root coefficient of [Eq. (1)] is a normalising constant which allows the comparison of different balances. A positive balance means that, in (geometric) mean, the group of parts in the numerator has more weight in the composition than the group in the denominator (and conversely for negative balances).

From the calculus of the robust RD_i distance and the inspection of the chi-square plot (here not reported), it was possible to clearly discriminate 901 anomalous compositions that resulted well separated from the remaining 3543 cases. Their relative position is reported in Fig. 2. As we can see, 3543 RD_i values (grey points) cover most part of the region, while outlier RD_i values (black points) mainly pertain to the coastal areas (saline intrusion) and to some zones where presence of minerali-

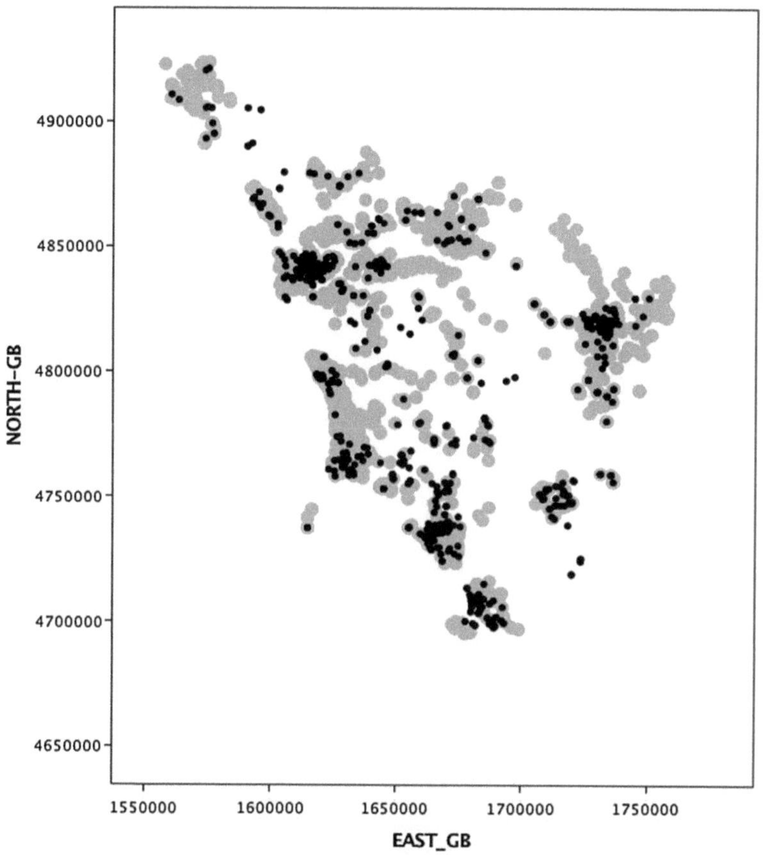

Fig. 2 Position of the anomalous groundwater samples (*black points*, $n = 901$) compared with the rest of the dataset (*grey points*, $n = 3543$) (methodology modified from [21]). GB = Gauss-Boaga coordinates reference system

sation, geothermal activities or pollution due to anthropic contributions can explain the presence of geochemical anomalies.

This first discrimination of the data allows us to identify anomalous compositions with a well defined geochemical explanation but does not assure that the remaining cases follow a multivariate normal distribution. In fact, the application of several statistical tests indicates that this hypothesis must be rejected ($p < 0.05$, MNV R-package, [31]). The chemical composition of the robust barycenter (obtained by back-transforming the ilr results of the analysis) in mg/L is given by $HCO_3 = 356$, $Ca = 102$, $Cl = 51$, $Mg = 21$, $K = 2.23$, $Na = 40.64$, $SO_4 = 55.51$, with a Total Dissolved Solids content (TDS) equal to 628 mg/L. Since its identification was not affected by anomalous values, it could represent a potential baseline hydrochemical facies, resulting as the most frequent facies for the investigated area.

As most groundwater that is used for water supply is derived from rainfall (TDS <20 mg/L, less in rural areas far from atmospheric pollution), its chemistry is a result of the chemistry of rainwater and the soils and rocks through which the water has passed, and the residence time in the aquifer. As residence time increases, groundwater tends to pick up more and more dissolved solids as a result of mineral dissolution (or weathering) reactions and its TDS value increases, according to the lithology of the areas [37]. From this point of view, the water associated with the lowest robust distance from the barycenter pertains to a typical Ca–HCO_3 geochemical facies (composition in mg/L equal to HCO_3 = 581, Ca = 168, Cl = 803, Mg = 32.7, K = 3.4, Na = 67.9, SO_4 = 97.6), increasing its TDS to 1754 mg/L, while the water with the higher distance pertains to a classical Ca–SO_4 geochemical facies (composition equal to HCO_3 = 4.8, Ca = 356, Cl = 270, Mg = 104, K = 2.4, Na = 250, SO_4 = 1250) with TDS equal to 2237 mg/L.

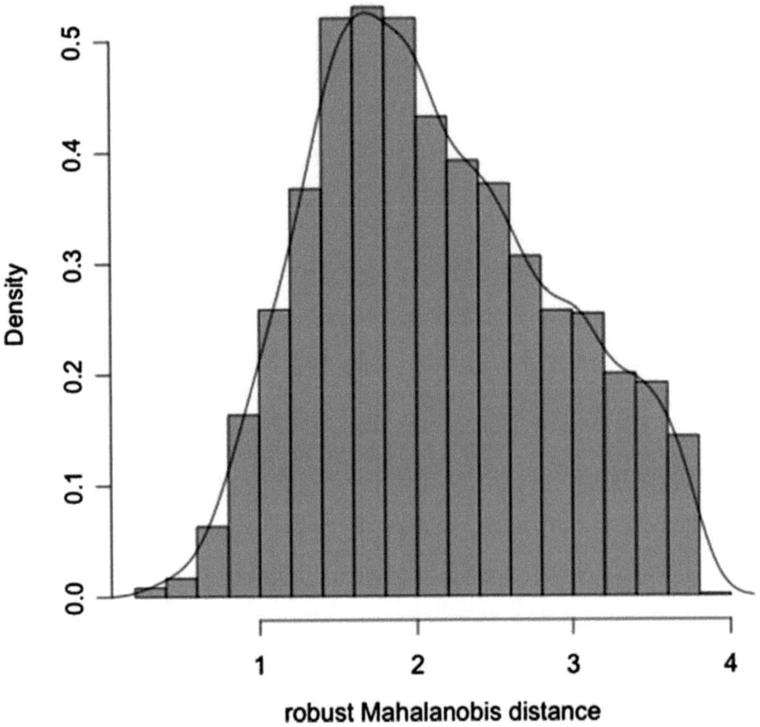

Fig. 3 Histogram of the Mahalanobis distance of n = 3543 groundwater samples after having eliminated 901 cases clearly separated from the others using the chi-square plot method [21]

3.2 The Distributional Behaviour of Robust Mahalanobis Distance

After having eliminated 901 cases from the whole dataset and verified that the remaining data distribution is not multivariate normal, interesting information can now be obtained by the analysis of the histogram of the robust Mahalanobis distance, as reported in Fig. 3. Its analysis is important to understand how compositions move from the barycentre or if data are again fragmented in sub-sets or if we are faced with a homogeneous set of data, even if not multivariate normal. Moreover, if the robust barycentre is supposed to represent a possible hydrochemical (compositional) baseline (the most frequent composition), the investigation of variability of compositional changes from this reference point could be monitored by the distributional behaviour of the robust Mahalanobis distance.

The RD_i data distribution appears to be slightly asymmetrical and apparently classical normal or lognormal models are not able to describe it. A model that leads towards much heavier tails is the power law distribution, often used to describe inhomogeneous and irregular distributions of element concentrations in several geological situations. When a probability density function displays a heavy non-Gaussian tail, this may be indication of the presence of multifractal processes [2]. This also indicates that the system as a whole is experiencing a non-linear dissipation in the energy interchange among different scales [23]. From this perspective the shape of a probability density function can be a powerful diagnostic tool in environmental problems. A nonnegative random variable X is said to have a power law distribution if:

$$Pr\left[X \leq x\right] \cong cx^{\alpha} \tag{3}$$

for constants $c > 0$ and $\alpha > 0$, so that asymptotically the tails fall according to the power α. For a power law distribution usually α falls in the range $0 < \alpha \leq 2$, in which case X has infinite variance. If $\alpha \leq 1$ then X also has infinite mean. The scaling exponent α is usually called the fractal dimension D. From a general point of view the model leads to much heavier tails than other common models explaining the concentration of minor and trace elements in geological materials [22, 25].

An interesting feature of this distribution model is that if X has a power law distribution, then in a log-log plot of $\Pr[X \leq x]$ the pattern of points will be described by a straight line. The graphical analysis is often used to visualise breaks in the data distribution (change in slope and presence of more than one straight line) and to investigate geochemical anomalies versus background [1]. Results for the studied groundwater data are reported in Fig. 4. As we can see a single straight line is not sufficient to describe all the data. Apparently, compositional changes that move compositions from the robust barycenter are multifractal and the physical-chemical reasons for this behaviour could be related to the presence of non-linear interactions between different scales and to the inhomogeneous character of dissipation of the chemical reactions.

In general, multifractality is a property of a dynamical system in which energy dissipation cannot be neglected. This condition leads to the presence of extended areas (or intervals) of low fluctuations intermittent with small areas of extremely large fluctuations. The spatial distribution of the RD_i values related to different segments of Fig. 4 is reported in Fig. 5.

As we can see, compositions similar to that of the robust barycenter are near to or in some cases overlap compositions very far from this reference, indicating that the chemistry of our samples can change in a short spatial/temporal range. Even if more research is needed in this direction, on the whole results suggest that the investigated water-rock system developed under conditions far from equilibrium as a progressively self-organising dissipative structure [24]. A "dissipative structure" is a non-equilibrium system, far from an equilibrium state, and should be supported by continuous inputs and outputs of materials in open conditions. Several authors have discussed the features of dissipative structures for groundwater flow systems and human activity appears to be a significant perturbation factor [39]. In this respect, the evolution of the investigated groundwater system may be attributable to a dissipative process of macroscopic states being perturbed by natural and human factors. Consequently the groundwater system is an open and complicated framework where interactions are governed by non-linear dynamics. In this context the identification of baseline hydrochemical compositions, as in our case, could be more representative of single values in the understanding of the processes working in the investigated area. However, to be noticed here is the possibility to also consider the value of single variables, as extracted from the compositional barycenter, for defined (legal) purposes. This makes the use of the concept of baseline compositions a practicable path.

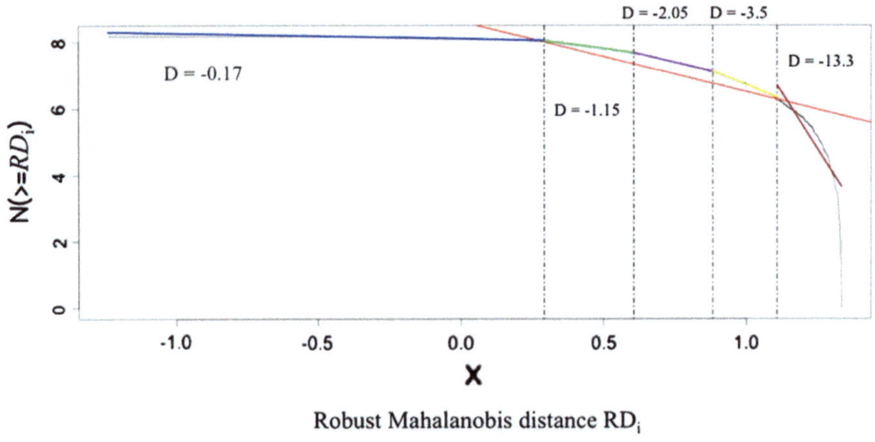

Fig. 4 Cumulative number N of samples whose robust Mahalanobis distance is equal to or higher than a given value RD_i on a log–log scale. D (slope of the segments) is the fractal dimension

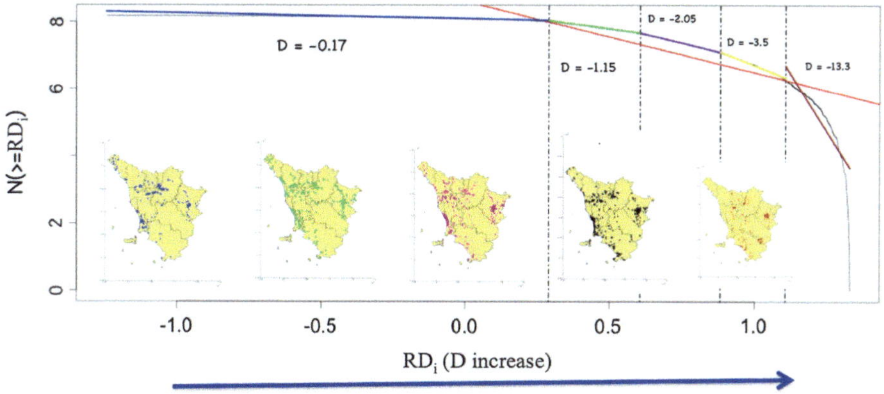

Fig. 5 Spatial distribution of the RD_i values partitioned according to the five segments of Fig. 4. Colours of the points on the maps correspond to the colors of the different segments

4 Conclusions

The term "geochemical baseline", officially introduced in 1993 in the context of the International Geological Correlation Program (IGCP Project 360), Global Geochemical Baselines, refers to the natural variation in the concentration of an element in the media of the superficial environment. The term can indicate the actual content of an element at a given point in time/space. It includes the geogenic natural concentrations (natural background) and the anthropogenic contribution.

Compositional changes from the robust barycenter for groundwater chemistry of Tuscany Region (central Italy) were analysed by investigating the behaviour of the robust Mahalanobis distance. The distance was calculated after having transformed original concentrations by using balances (a particular type of isometric log-ratio transformation, [5]). The aim was to verify (1) if the robust barycenter could be considered a possible baseline composition, (2) which type of information was contained in the variability of the robust distance values. After having eliminated the more anomalous compositions from the whole dataset [21], the distributional form of the robust Mahalanobis distance was investigated. The application of normality tests indicated that the hypothesis of multivariate normality cannot be accepted and that the distance obeys the power law distribution, displaying properties of multifractality. In this perspective, compositional changes from barycenter are neither completely deterministic not totally chaotic. Rather they are in an intermediate state, which possesses a property of multifractality in the spatial domain and probable intermittency in time.

Since fractal structures form spontaneously only in the presence of a complex dissipative structure, the investigated groundwater system appears to be an open and complicated framework where interactions are governed by non-linear dynamics. Moreover, compositional changes are characterised by self-similarity patterns. This means that the physical–chemical laws that control spatial and temporal variability on one scale also control patterns and spatial variability on other scales, implying scale-independence.

All of the previous results, even if preliminary, require caution in the definition of baseline concentrations for single values without the use of a compositional approach and a global conceptual model of the groundwater system for all its interconnected components.

Acknowledgments The research was developed thanks to funds of the University of Florence (2014) and of Tuscany Region through the Geobasi project (2009–2014). Santiago Thio Fernandez de Henestrosa is thanked for the editing of the manuscript, Vera Pawlowsky-Glahn and an anonymous referee for their valuable suggestions, Elizabeth Hancock for the English language.

References

1. Agterberg, F.P.: Geomathematics: Theoretical Foundations, Applications and Future Developments.Springer Series in Quantitative Geology and Geostatistics, vol. 18 (2014)
2. Agterberg, F.P.: Mixtures of multiplicative cascade models in geochemistry. Nonlinear Process. Geophys. **14**, 201–209 (2007)
3. Aitchison, J.: The Statistical Analysis of Compositional Data (Reprinted in 2003 by The Blackburn Press), p. 416. Chapman & Hall Ltd., London (UK) (1986)
4. Aitchison, J.: The statistical analysis of compositional data (with discussion). J. Roy. Stat. Soc. Ser. B-Stat. Methodol. **44**(2), 139–177 (1982)
5. Buccianti, A., Egozcue, J.J., Pawlowsky-Glahn, V.: Variation diagrams to statistically model the behaviour of geochemical variables: theory and applications. J. Hydrol. **519(PA)**, 988–998 (2014)
6. Buccianti, A., Magli, R.: Metric concepts and implications in describing compositional changes for world rivers water chemistry. Comput. Geosci. **37**(5), 670–676 (2011)
7. Buccianti, A.: Is compositional data analysis a way to see beyond the illusion?. Comput. Geosci. **50**, 165–173 (2013)
8. Buccianti, A., Gallo, M.: Weighted principal component analysis for compositional data: application example for the water chemistry of the Arno river (Tuscany, central Italy). Environmetrics **24**, 269–277 (2013)
9. Buccianti, A., Grunsky, E.: Compositional data analysis in geochemistry: are we sure to see what really occurs during natural processes? J. Geochem. Explor. **14**, 1–5 (2014)
10. Carmignani L., Conti P., Cornamusini G., Meccheri M.: The internal northern Apennines, the northern tyrrhenian sea and the Sardinia-Corsica block. In: Geology of Italy. Italian Geological Society Bulletin IGC32 Florence-2004, pp. 59–77 (2004)
11. Daszykowski, M., Kaczmarek, K., Vander Heyden, Y., Walczak, B.: Robust statistic in data analysis. A review. Basic concept. Chemometr. Intell. Lab. Syst. **85**, 203–219 (2007)
12. De Caritat, P., Grunsky, E.: Defining element associations and inferring geological processes from total element concentrations in Australia catchment outlet sediments: Multivariate analysis of continental-scale geochemical data. Appl. Geochem. **33**, 104–126 (2013)
13. Egozcue, J.J., Pawlowsky-Glahn, V.: Simplicial geometry for compositional data. In: Buccianti, A., Mateu-Figueras, G., Pawlowsky-Glahn, V. (eds.) Compositional Data Analysis in the Geosciences: From Theory to Practice. Special Publication, vol. 264, pp. 12–28. Geological Society, London (2006)
14. Egozcue, J.J., Pawlowsky-Glahn, V.: Groups of parts and their balances in compositional data analysis. Math. Geol. **37**(7), 795–828 (2005)

15. Egozcue, J.J., Pawlowsky-Glahn, V., Mateu-Figueras, G., Barcelo-Vidal, C.: Isometric logratio transformations for compositional data analysis. Math. Geol. **35**(3), 279–300 (2003)
16. Filzmoser, P., Hron, K., Reimann, C.: Interpretation of multivariate outliers for compositional data. Comput. Geosci. **39**, 77–85 (2012)
17. Filzmoser, P., Hron, K.: Outlier detection for compositional data using robust methods. Math. Geosci. **40**(3), 233–248 (2008)
18. Filzmoser, P., Hron, K., Reimann, C.: Univariate statistical analysis of environmental (compositional) data: problems and possibilities. Sci. Total Environ. **407**, 6100–6108 (2009)
19. Filzmoser, P., Hron, K., Reimann, C.: Principal component analysis for compositional data with outliers. Environmetrics **20**(6), 621–632 (2009)
20. Galuszka, A.: A review of geochemical background concepts and an example using data from Poland. Environ. Geol. **52**, 861–870 (2007)
21. Garrett, R.G.: The chi-square plot: a tool for multivariate outlier recognition. J. Geochem. Explor. **32**, 319–341 (1989)
22. Goncalves, M.A.: Characteriszation of geochemical distributions using multifractals models. Math. Geosci. **33**, 41–61 (2001)
23. Hunt, A.G., Ghanbarian, B., Skinner, T.E., Ewing, R.P.: Scaling of geochemical reaction rates via advective solute transport. Chaos **25**(075403), 1–15 (2015)
24. Kondepudi, D., Prigogine, I.: Modern Thermodynamics. From Heat Engines to Dissipative Structures. Wiley (1998)
25. Ma, T., Li, C., Lu, Z.: Estimating the average concentration of minor and trace elements in surficial sediments using fractal methods. J. Geochem. Explor. **139**, 207–216 (2014)
26. Maronna, R.A., Zamar, R.H.: Robust multivariate estimates for highdimensional datasets. Technometrics **44**, 307–317 (2002)
27. Matschullat, J., Ottenstein, R., Reimann, C.: Geochemical background can we calculate it? Environ. Earth Sci. **39**(9), 990–1000 (2000)
28. Nieto, P., Custodio, E., Manzano, M.: Baseline groundwater quality: a European approach. Environ. Sci. Policy **8**, 399–409 (2005)
29. Nisi, B., Buccianti, A., Raco, B., Battaglini, R.: Analysis of complex regional databases and their support in the identification of background/baseline compositional facies in groundwater investigation: developments and application examples. J. Geochem. Explor. **164**, 3–17 (2016)
30. Nordstrom, D.K.: Baseline and premining geochemical characterization of mined sites. Appl. Geochem. **57**, 17–34 (2015)
31. R Development Core Team: R: A Language and Environment for Statistical Computing. R Foundation for Statistical Computing, Vienna, Austria. ISBN 3-900051-07-0 (2015). http://www.R-project.org/
32. Raco, B., Buccianti, A., Corongiu, M., Lavorini, G., Macera, P., Manetti, F., Mari, R., Masetti, G., Menichetti, S., Nisi, B., Protano, G., Romanelli, S.: The geochemical database of Tuscany Region (Italy). Ital. J. Groundwater **AS12055**, 007–018 (2015). doi:10.7343/AS-100-15-0127
33. Reimann, C., Garrett, R.G.: Geochemical background concept and reality. Sci. Total Environ. **350**, 12–27 (2005)
34. Rousseeuw, P.J.: Least median of squares regression. J. Am. Stat. Assoc. **79**, 871–880 (1984)
35. UN-Water: Annual report: Available via DIALOG. http://www.unwater.org (2011). Accessed 15 Jan 2016
36. Verboven, S., Hubert, M.: LIBRA: a MATLAB libray for robust analysis. Chemometr. Intell. Lab. Syst. **75**, 127–136 (2005)
37. West, L.J., Odling, N.E.: Groundwater. In: Holden, J. (ed.) Water Resources. An Integrated Approach. Routledge, Taylor & Francis Group (2014)
38. World Health Organisation (WHO): Our plane, our health. Report of WHO Commission on Health and Environment, Geneva, World Health Organisaton (1992)
39. Xu, W., Du, S.: Information entropy evolution for groundwater flow system: a case study of artificial recharge in Shijiazhuang city, China. Entropy **16**, 4408–4419 (2014)

Multielement Geochemical Modelling for Mine Planning: Case Study from an Epithermal Gold Deposit

N. Caciagli

Abstract Mineralisation and alteration processes will result in zones with distinct geochemical characteristics within an orebody. To visualise the mine scale variability that arises as a result of these processes, geochemical domains are defined using a k-means clustering algorithm to analyse multielement data. The fact that the chemical values can be grouped in a defined 3D location clearly suggests that the clusters have meaning in terms of geological process. Principal component analysis (PCA) of these clusters can further improve the understanding of this variability. These clusters form the basis of the geochemical domains which have direct implications for characterization and proportional sampling of geometallurgical and waste rock domains. In the case study presented, pre-mining geochemical characterisations were undertaken at an epithermal gold deposit to support metallurgical sampling and mine planning. k-means cluster analysis and principal component analysis of the geochemical clusters was used to support the metallurgical sampling programme by identifying domains for variability testing. The geochemical clusters identified were used to define the oxide, sulphide and transition zones, a critical factor for mineral processing and recoveries and a key variable in the economics of the project. The R software environment for statistical computing was used for exploratory data analysis (e.g. PCA; zCompositions, robCompositions) and k-means analysis (fpc).

Keywords Compositional data analysis · PCA · Cluster analysis geology · Mining

1 Introduction

A new gold or base metal mine can take 10–15 years to permit and construct and involves an investment of >$1B USD. Inadequate characterization of the orebody can lead to unexpected issues with geotechnical stability, metallurgical recoveries, environmental impacts, or other problems. Kinross has determined that a thorough

N. Caciagli (✉)
Kinross Gold Corporation, Toronto, ON, Canada
e-mail: Natalie.caciagli@Kinross.com

J.A. Martín-Fernández and S. Thió-Henestrosa (eds.), *Compositional Data Analysis*, Springer Proceedings in Mathematics & Statistics 187,
DOI 10.1007/978-3-319-44811-4_4

understanding of its orebodies is critical to the success of these major, long-term investments and that geochemistry is a key component of this understanding.

Mineralizing and alteration processes will result in zones with distinct geochemical characteristics. The fact that the chemical values can be grouped in a defined 3D location clearly suggests that the clusters have meaning in terms of geological process. This forms the basis of the geochemical domains, which have direct implications for sampling locations for variability testing of geometallurgical recoveries.

The geochemical domains will reflect the mineralogy of the domains with a granularity not always visible by geological mapping and logging alone. For example, alteration events associated with gold mineralization can overprint the host lithology to a degree that the original rock type is unable to be properly identified, or metasomatism may result in chemical changes to mineralogy not fully evident through visual inspection alone. The use of geochemical domains supports a more concrete link between mineralogy and metallurgical variables such as cyanide consumption, metal recoveries, concentration of deleterious elements, as well as comminution properties (hardness, grindability).

2 Methodology

The ultimate goal of the geochemical modelling process is to produce geochemical domains that are relevant for metallurgical study on a scale compatible with the proposed mining method. As such there will be a trade-off between granularity and complexity of the model. Geochemical domains need to be viewed within the context of lithology and alteration so that metallurgical variables can be linked to mineralogy.

2.1 Data Pre-processing

The dataset used for the geochemical modelling must consist of spatially colocated data. For example; samples for gold analysis and multielements analysis should be collected over the same intervals. Any hyperspectral logging (qualitative mineralogy) or QEMSCAN (quantitative mineralogy) should also be spatially colocated with the geochemical and metal assay data. The sampling strategy needs to be assessed for possible sampling bias. This can be accomplished by visualisation of the data in 3-D space, examination of sample lengths and counts by lithology, alteration, structure through histograms and box and whisker diagrams. Exploratory data analysis of the multielement geochemistry (i.e. summary stats, detection limits) is key to understanding the nature and distribution of the dataset [4].

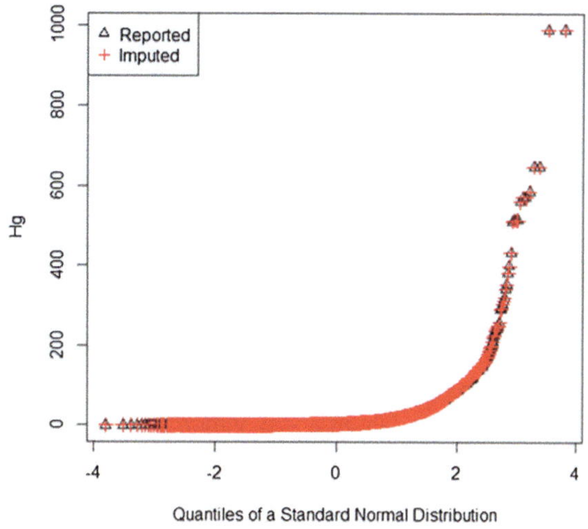

Fig. 1 Quantile–quantile plot of imputed values and reported values for Hg

2.1.1 Data Imputation

The modified "Expectation-Maximisation" (EM) algorithm from the zCompositions package [6] within the R statistical computing programme [7] was used to impute values below the lower limits of detection (LOD) as well as any missing values in the dataset. This procedure fills in the lower tail of the distribution of each element while still preserving the covariance structure of the data. The zCompositions modified EM-imputation algorithm imputes values based on a reference dataset (i.e. samples with no missing observations) and on the threshold values (i.e. the limit of detection for a given element). Following the application of the imputation routine, every imputed value was visually inspected on a quantile–quantile plot (Fig. 1) for a given element to assess the appropriateness of the replacement.

2.1.2 Data Transformation

The working dataset was transformed using log ratio normalisation [1]. Compositional data is by definition constrained to a constant sum (e.g. reported as percent or per mil) and as such the individual variables will not vary independently. This induced correlation introduces a bias into the covariance structure of the data and may obscure the true relationships between the variables. Log ratio normalisation takes into account the constant sum constraint of compositional data and centres the data in a way that removes this bias. For robust PCA (see below) the working datasets are normalised using an isometric log ratio transform (ilr; Egozcue et al. [2]). k-means clustering is performed on a centred log ratio transformed (clr) dataset.

2.2 rPCA

In a mining environment, understanding the behaviour of outliers is of significant interest. Gold mineralization tends to be inherently non-uniform and as a result many samples within the ore zone may have "anomalous" values compared to the waste rock. These samples need to be understood with respect to their geological context (lithology, alteration and structure) for vectoring and targeting purposes, as well as for understanding the nature of the gold distribution within the ore body. Unless sampling issues or analytical errors are confirmed the dataset is examined inclusive of outliers.

Principal component analysis (PCA) is the simplest of the true eigenvector-based multivariate analyses. Its purpose can often be thought of as revealing the internal structure of the data in a way which best explains the data variance. It is a technique used to change a set of original variables into a number of basic dimensions. The algorithm used here for the calculation of the principal components was the "Robust PCA" from the robCompositions package [3, 8, 9] within the R statistical computing programme. Robust PCA (rPCA) is less affected by outliers in the dataset and provides more reliable calculation of the covariance matrix for the analysis. The rPCA algorithm from the robCompositions package utilises a minimum covariance determinant (MCD) by examining a subset of h observations with the smallest determinant of their sample covariance determinant. In order to maximise the robustness of the MCD location, the h used here was 1/2 of the total sample size because the number of outliers tends to be high.

2.3 k-Means Cluster Analysis

k-means clustering is a method of cluster analysis that aims to partition n observations into k clusters, in which each observation belongs to the cluster with the nearest mean. k-means clustering uses the minimum Euclidian distance (difference between values) as the main criterion to discriminate between different groups. These clusters can be back-coded into the database and viewed in 3D space for interpretation within the context of lithology, alteration and metallurgical responses.

The number of clusters to partition a dataset can be assessed using a plot of the sum of squared errors (SSE) versus number of clusters (Fig. 2). The best number of clusters to describe the structure in the dataset can be determined by locating the break in slope on a plot of the SSE versus number of clusters. For geochemical modelling of an orebody a combination of the SSE plot, visualisation of the clusters in either principal component space or classical discriminant coordinates and a visual (3D)

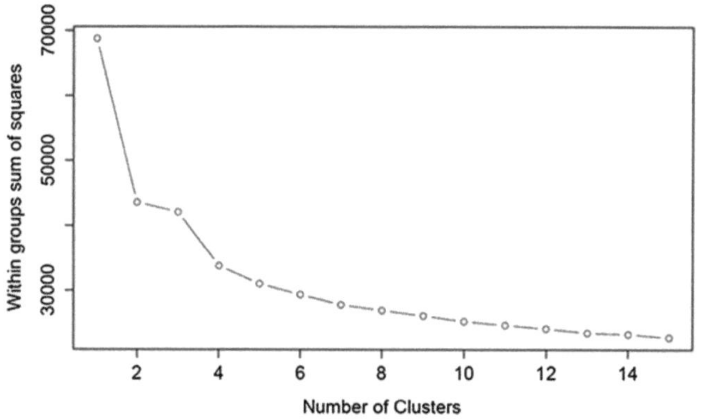

Fig. 2 An example of a sum of squared errors (SSE) plot versus number of clusters

inspection of the clusters in a geological modelling software was used to determine the "best" number of clusters. For example, from Fig. 2, four clusters are suggested; however five clusters showed distinct separation in 3-D space (see Fig. 4).

2.3.1 Statistical Validation of Clusters

The robustness of the clusters was determined using a bootstrapping algorithm from the fcp package [5] within the R statistical computing programme that resamples the dataset 100 times and returns a measure of the stability of the cluster, the Jaccard similarity value, which is assigned a value between 0 and 1. Generally, a valid, stable cluster should yield a mean Jaccard similarity value of 0.75 or more [5].

2.3.2 Geochemical and Spatial Validation of Clusters

The geochemical variability of the clusters can be examined within either principal component space or with a discriminant analysis. The sample dataset was tagged with the cluster number and plotted with their rPCA coordinates (see Fig. 3). This type of plot can also provide as assessment as to the suitability of the number of clusters selected for k-means cluster analysis. For example, samples from cluster 7 (red) and cluster 5 (blue) completely overlap in PC1–PC2 space, suggesting that there is very little variability between these clusters. These samples were then examined spatially to determine if they have a defined location in space (Fig. 4) and can be explained within the context of known lithology, alteration or structure or whether they just describe the inherent geochemical heterogeneity within that domain.

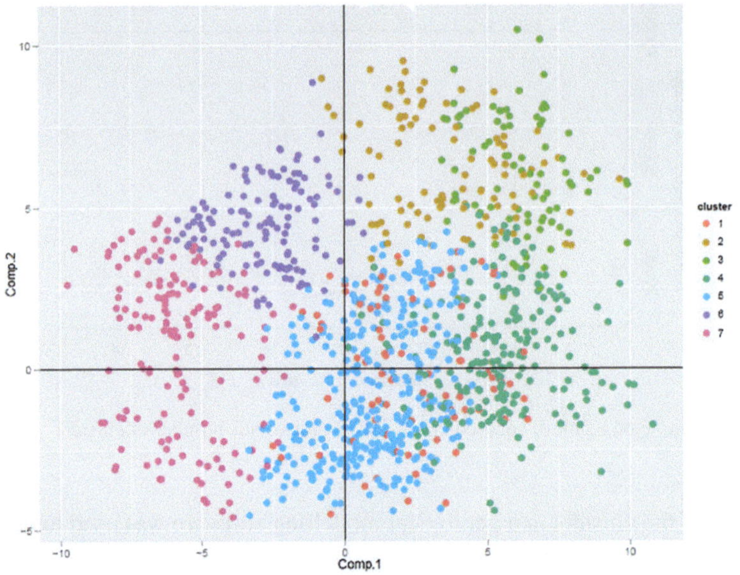

Fig. 3 Plot of PC1 versus PC2 coordinates for samples colour coded by cluster number

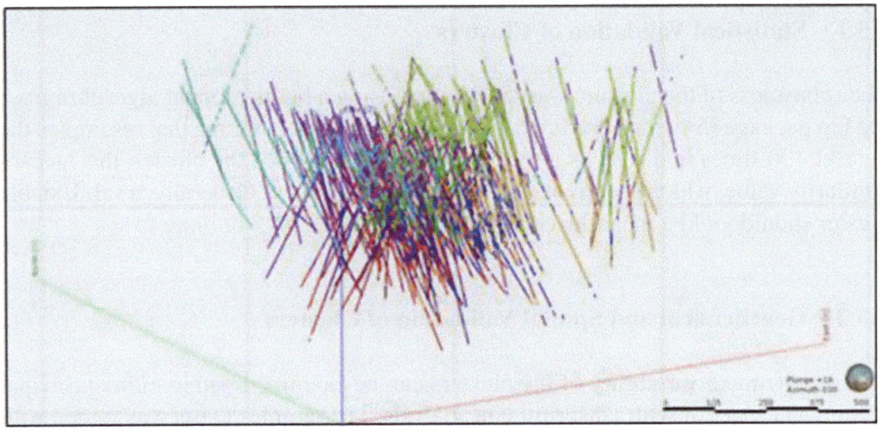

Fig. 4 Drill hole intervals back-coded with cluster number viewed in 3-D space

2.4 *Wireframes and Characterization of Domains*

The first pass wireframe construction process uses Leapfrog™ geological modelling software to create surfaces based on the criteria outlined above to encapsulate the back-coded clusters. The surfaces are created independently of the lithological or alteration domains, however the geochemical domains will correspond to a combination of the lithological, alteration and weathering domains determined from visual logging.

3 Case Study

Deposit A is interpreted as having been emplaced into a phreatomagmatic diatreme–dome complex with local diatreme fill. Mineralization is associated with high sulphidation vuggy silica with advanced argillic quartz–alunite or quartz–pyrophyllite alteration, which grades outward into quartz-clays. The alteration assemblage appears to have a high percentage of silica and lesser amounts of clay. In oxidised portions, the gold and silver are residual, but at depth gold and silver are associated with multiple sulphides (mostly enargite and pyrite, with lesser covellite and sphalerite) and sulphosalts.

The depth of oxidation varies from less than 100 m to over 300 m and is typically around 120–150 m. There is a narrow and discontinuous "mixed" horizon at the oxide–sulphide interface (transition zone), less than 10 m thick, but it invariably includes secondary chalcocite.

Within this transition zone, metal recoveries can be significantly less than in the oxide zone; in this case, 75 % recovered Au compared to 82 %. A robust definition and identification of this domain is critical to determining the economics of the project. However determination of this transition zone by visual logging is difficult and subjective.

A detailed geochemical model was created within the mineralised vuggy silica zone of Deposit A to determine a consistent and unbiased definition of the oxide, transition and sulphide zones to accurately assess the amount of recoverable metals and better understand the economic potential of the deposit.

3.1 Data and Data Pre-Processing

The working dataset consisted of information collected on two very different resolutions. The logged lithology, logged alteration and metal assay data were collected with a 2 m resolution. In contrast, the multielement data (collected by previous operators) consisted of 10 m composites created by taking a 20 g sample from every 5 assay sample pulps (originally collected on 2 m intervals) down hole, homogenising those 5 samples and sending an aliquot of the composite for aqua regia digest and multielement analysis by ICP. In order to properly merge the two datasets the data

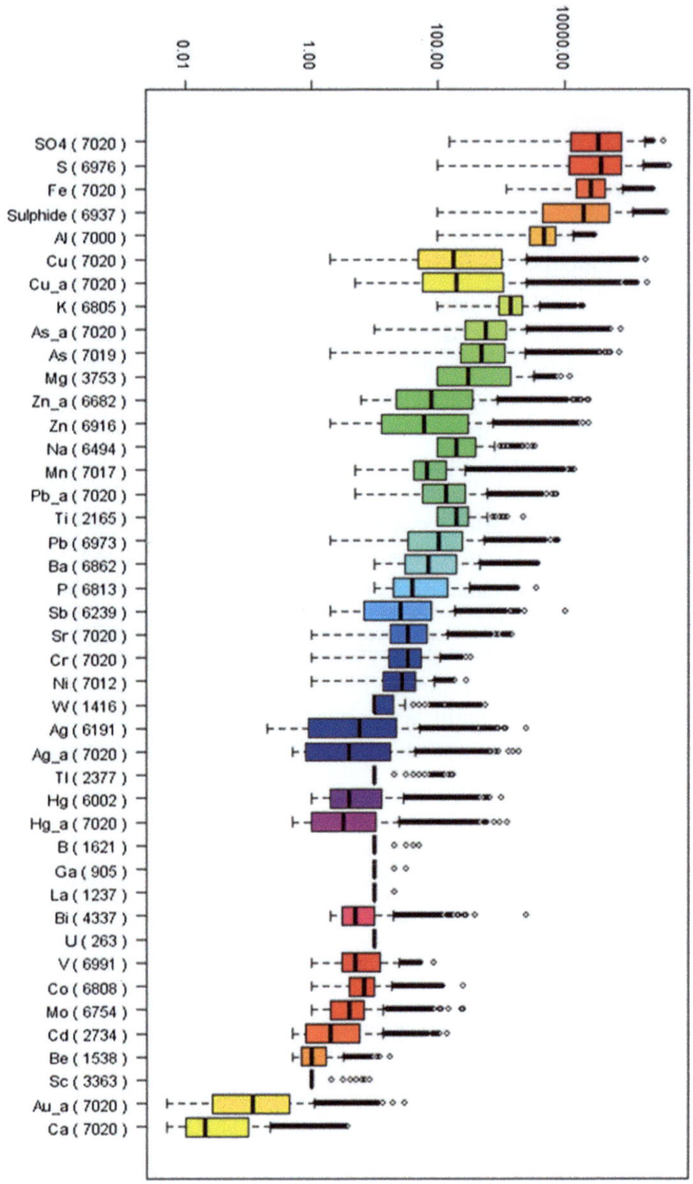

Fig. 5 Elemental distribution of raw data—all expressed as ppm, (n) indicates samples > LOD

in the 2 m Au, Cu, Ag, As, Pb, Zn and Hg assay tables needed to be composited over 10 m. Multielement data sampled over 2 m intervals, consistent with the Au sampling protocol would be preferable.

The statistical distribution of each element was inspected individually. Elements were removed from the dataset if a significant portion of samples measured below detection limits (>50 %) or if the data was severely quantized due to the analytical resolution. This resulted in the removal of Sc, Cd, Tl, Ti, Th, B, Be, W, La, Ga and U (Fig. 5).

3.2 *rPCA*

The Deposit A rPCA was constructed using centre log ratio transformed multielement data to examine relationships found in Au-bearing lithology and alteration types. The numerical variables analysed were the assayed elements: Au, Ag, As, Al, Ba, Bi, Ca, Co, Cr, Cu, Fe, Hg, K, Mg, Mn, Mo, Na, Ni, P, Pb, Sb, Sr, S-total, Sulphide, Sulphate, V and Zn.

Partial digestion, using aqua regia, was used to extract elements for analysis. This method dissolves minerals selectively and therefore the bulk rock chemistry is not reflected in the major elements determined by this method. The recovery of each element will depend on the mineralogy. The aqua regia method only targets the trioctahedral silicates (biotite, chlorite), clays, sulfosalts and some oxides. It will not recover elements hosted by other silicates and refractory oxides. Consequently, extra care must be given to interpretation of the major element data obtained by partial rock dissolution.

The observed chemical relationships of selected elements show patterns that are assumed to be a function of the type of mineralization and accompanying alteration. In the rPCA variable map in Fig. 6, a number of elements are correlating with Au (Cu, Ag, Hg, Fe, SO_4 and S; the vectors are in close proximity to each other). The presence of enargite (Cu_3AsS_4) is suggested by the Cu–As association. The strong correlation of Al–K implies an illite (K–Al clay) association. The vector describing Zn is the longest vector in the PCA suggesting that the occurrence of sphalerite (ZnS) is highly variable; however as demonstrated in the k-mean analysis, discussed below, the Zn variability can be spatially defined.

3.3 *k-means cluster analysis*

All the elements used in the rPCA are also used in the clustering exercise. For Deposit A, the preliminary cluster analysis resulted in domains that broadly correspond to the alteration zones (see Fig. 4). To provide further resolution of the variability within only the mineralized ore zone, a k-means cluster analysis was carried out on the geochemical data from intervals in the vuggy silica oxide, vuggy silica sulphide and

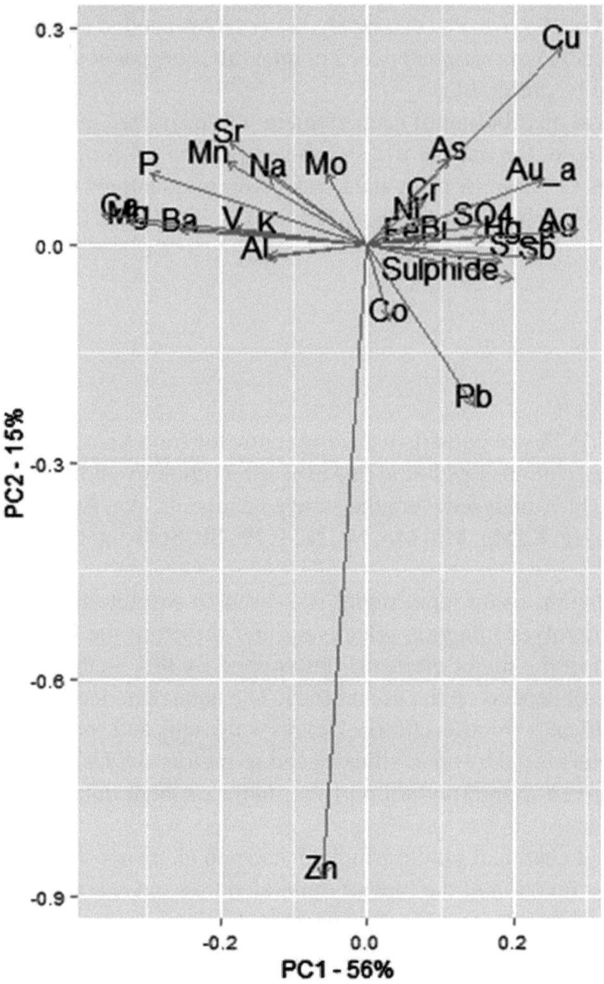

Fig. 6 rPCA variables map for deposit a multielement data

alunite alteration zone. In this deposit, the geochemical contrast is so high between the various alteration zones and between the fresh and weathered zones that the *k*-means clustering can be applied to the clr-transformed variables. On other cases applying the *k*-means clustering can be improved, cancelling much of the "noise" and focusing on specific processes, by clustering specific principal components.

For Deposit A, a combination of the SSE plot, visualisation of the clusters in discriminant coordinate space (Fig. 8) and 3D space (Fig. 4) was used to determine the appropriate number of clusters.

From Fig. 7, four clusters are suggested; however, five clusters showed distinct separation in both geochemical space (Fig. 8) and 3D space (Fig. 4).

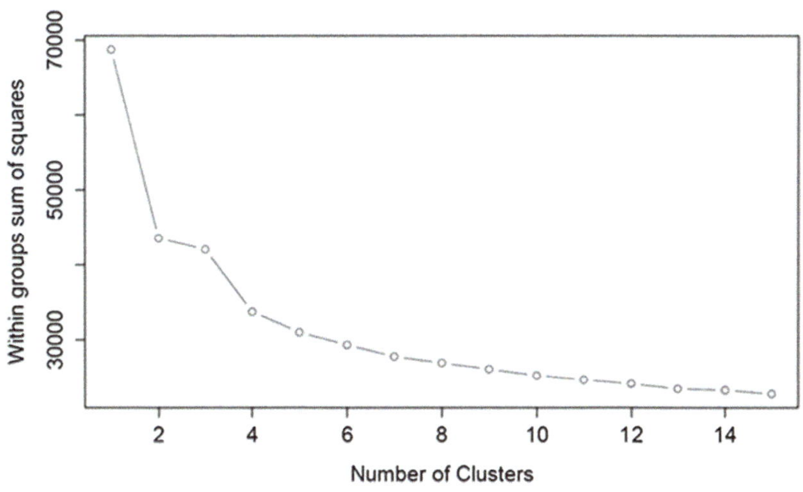

Fig. 7 Sum of squared errors (SSE) versus number of clusters

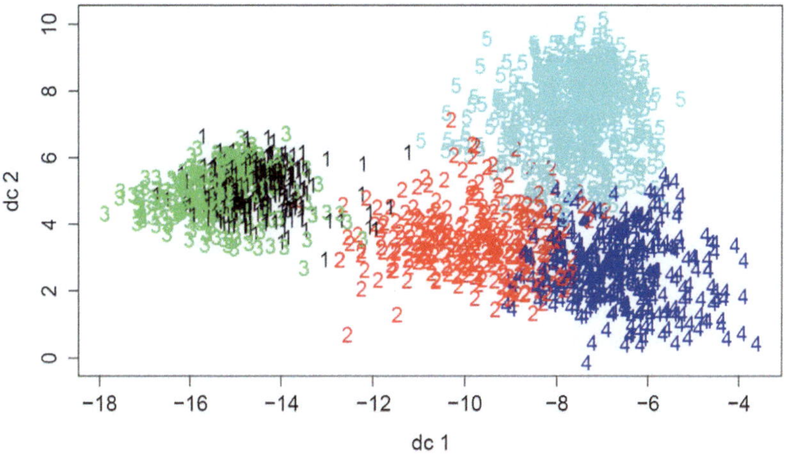

Fig. 8 Visualisation of the clusters in geochemical space

The robustness of the clusters was determined using a bootstrapping algorithm which resamples the dataset 100 times and returns the Jaccard values, a measure of the stability, for each cluster. The mean Jaccard values for Clusters 2, 4 and 5 range from 0.82 to 0.84, while the Jaccard values for Cluster 1 and Cluster 3 are 0.66 (Figs. 4 and 8).

These clusters can be briefly characterised as

Cluster 1: A low copper (mean content of 0.5 %) high zinc domain within the sulphide zone corresponding to sphalerite rich zones. The gold within this domain is highly refractory and not recoverable by current processing methods.

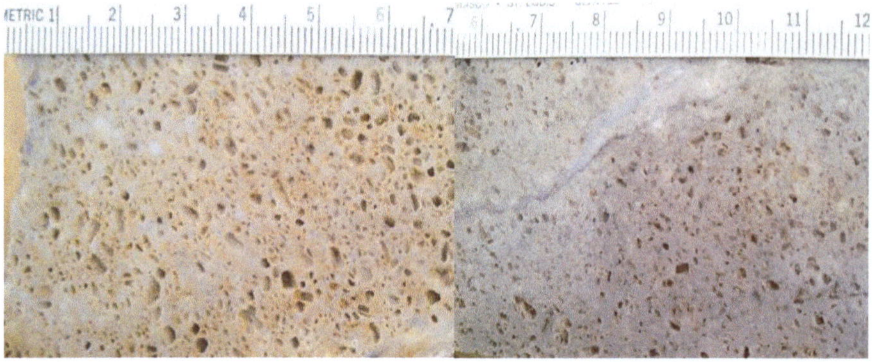

Fig. 9 Drill core photos of intervals classified as oxide domain (cluster 4) and transition domain (cluster 2). Within the oxide zone vugs are either empty or filled with quartz (A). Within the transition domain, vugs are filled with quartz + fine grained disseminated sulphides (*photo* is cm scale)

Cluster 2: A "mixed" horizon at the oxide-sulphide interface with elevated soluble copper content corresponding to the transition domain.

Cluster 3: A high copper (mean content of 2 %) and arsenic domain within the sulphide zone corresponding to enargite rich zones. Due to the high arsenic content and the refractory nature of the gold in this domain is not economically recoverable by current processing methods, however due to the higher Cu content this zone could be significant.

Cluster 4: An oxide domain (mean S content of 0.2 %) within the vuggy silica zone corresponding to the oxide zone. This domain has the highest gold recoveries.

Cluster 5: A potassium and aluminium rich domain that slightly overprints the vuggy silica alteration and transitions into the quartz–illite alteration halo on the margin of the deposit. There is recoverable metal in this domain, but the differing mineralogy will result in differing behaviour in the mill. This domain also contains the highest mean Pb content of all the domains.

Cluster 2 (transition domain) and 4 (oxide domain) occur predominantly within the portion of the deposit logged as oxide. Visual logging frequently misidentifies transition and oxide materials, particularly near the oxide-sulphide interface due to difficulty in visually identifying these zone and subjective nature of the geological logging (Fig. 9). Because the identification of the transition zone is relevant for metallurgical processing and key to determining metal recoveries, this geochemical determination provides an unbiased and consistent way to identify this zone.
The transition domain at the oxide–sulphide interface includes the secondary copper minerals, chalcocite. Chalcocite provides high concentrations of soluble copper will have a negative effect on the total cyanide consumption and possibly reduced gold recoveries.

Figure 10 shows how the spatial separation of clusters 2 and 4 relate to the redox limits as established by visual logging. Chemical cluster 4, interpreted as the oxide

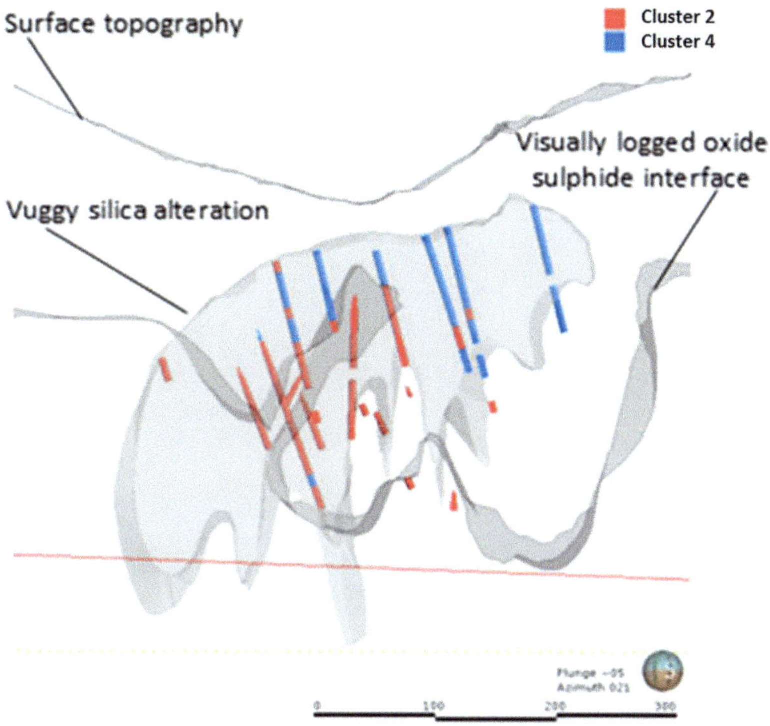

Fig. 10 A vertical cross section through the ore zone showing the lithological and spatial relationship of the geochemically determined oxide zone (*blue* intervals) and transition zone (*red* intervals). Also shown is the logged vuggy silica domain (*light grey* shape) and the redox boundary identified by visual logging (*darker grey*)

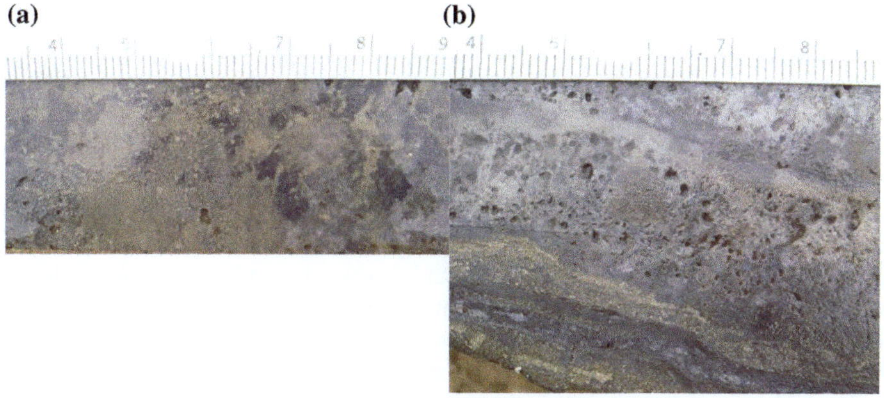

Fig. 11 Drill core photos of intervals classified as **a** low Cu–high Zn sulphide domain (cluster 1) characterised by coarse-grained interlocking sulphides and **b** high Cu–As sulphide domain (cluster 3) characterized by massive Cu-rich sulphide veins (*Photo* is cm scale)

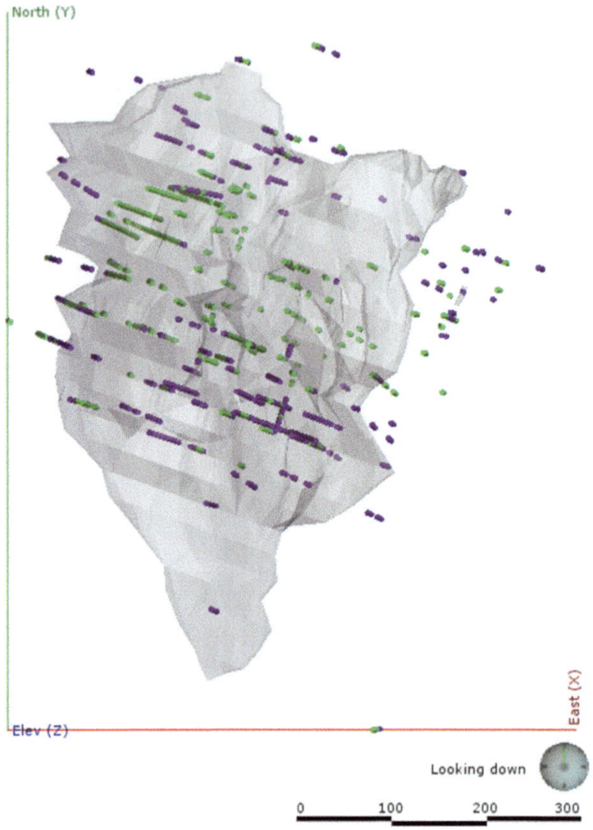

Fig. 12 Plan view of the logged vuggy silica alteration (*grey shape*) within the sulphide zone showing the spatial relationship between the high Cu–As sulphide domain (cluster 3; *green* drill strings) and the low Cu–high Zn sulphide domain (cluster 1; *purple* drill strings)

domain, is hosted by the vuggy silica alteration well within the visually logged oxide zone while the intervals assigned to cluster 2, interpreted as transition domain, correspond to the vuggy silica alteration at the interface between the oxide and sulphide zones.

Clusters 1 and 3 occur within the visually logged sulphide zone and differ from each other in copper content. Cluster 3 has mean copper content of 2 % and cluster 1 has a mean copper content of 0.5 %. Cluster 3 most likely represents copper rich feeder zones or copper rich veins within the low copper domain (cluster 1; Fig. 11). However, the overlap between cluster 1 and 3, spatially and geochemically likely reflects the inability of the 10 m composited data to precisely discriminate at this scale. Depending on the ultimate metallurgical processing flowsheet, resolution of these high copper zones may or may not be necessary.

The high copper domain within the sulphide zone of vuggy silica alteration (cluster 3) is centred on the main feeder zones of the vuggy silica alteration. The low copper domain (cluster 1) is more predominant on the peripheries of the vuggy silica alteration zone to the north and south (Fig. 12).

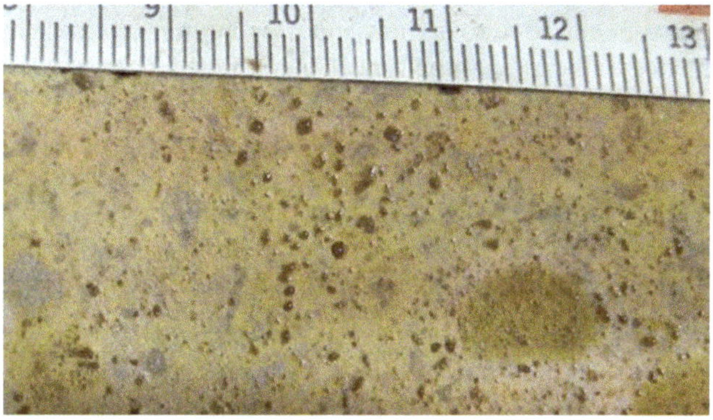

Fig. 13 Drill core photo of interval alunite-rich domain (cluster 5). These rocks are easily identified by their distinct yellow-beige colour

Fig. 14 Plan view of the logged vuggy silica zone (*grey*) and logged advanced argillic alteration (*yellow*) showing the spatial relationship between the alunite-rich domain (*light blue* drill strings) and the logged geology and alteration

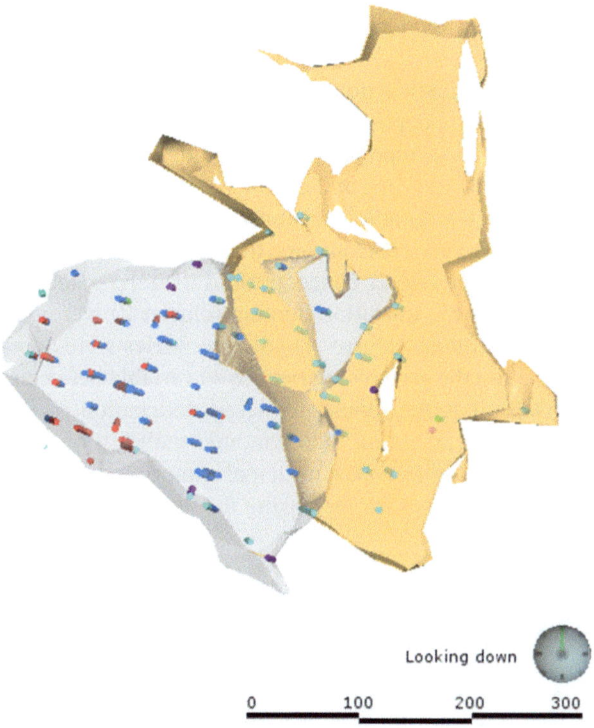

Cluster 5 corresponds to a potassium and aluminium rich domain that slightly overprints the vuggy silica alteration on the margin of the deposit. This domain is characterized by high sulphate content consistent with the appearance of the sulphate minerals alunite and jarosite (Fig. 13). This domain extends beyond the logged vuggy silica zone into the advanced argillic alteration halo within the host rock (see Fig. 14). Due to the presence of sulphate minerals, the ore in this domain has a higher lime consumption and is slightly softer than the vuggy silica ore. Characterization of this domain is relevant for mine planning because these variables can affect processing costs and throughput estimates.

4 Conclusions

Geochemical domains were created using a k-means clustering algorithm to analyse multielement data from the vuggy silica zone of a high sulphidation epithermal gold deposit (Deposit A). This analysis was able to partition the deposit into several distinct geochemical domains that correlate with the known geological and alteration domains of the deposit, but also provided additional information on the delineation of the transition zone between the oxide and sulphide zones.

The transition zone at the oxide-sulphide interface includes secondary chalcocite, Cu_2S, which is soluble in cyanide solutions. When cyanide consumption and copper gold recoveries were examined the transition domain (cluster 2) was found to contain higher soluble copper concentrations than the oxide domain (cluster 4). Gold recoveries within the transition domain (cluster 2) were on the average of 75 %. Within the oxide domain (cluster 4) gold recoveries were approximately 83 %.

Because the gold recoveries are so variable between the oxide and transition zone for Deposit A the robust, unbiased delineation of the oxide and transition zones is critical to assessing the economic viability of this project. Compositional data analysis, including robust principal component analysis and k-means cluster analysis is an effective tool for increasing the understanding and knowledge of an orebody.

References

1. Aitchison, J.: The Statistical Analysis of Compositional Data (Reprinted in 2003 by The Blackburn Press), p. 416. Chapman & Hall Ltd., London (UK) (1986)
2. Egozcue, J.J., et al.: Isometric log ratio transformations for compositional data analysis. Math. Geology. **35**(3), 279–300 (2003)
3. Filmozer, P., Hron, K.: Robust statistical analysis. In: Pawlowsky-Glahn, V., Buccianti, A. (eds.) Compositional Data Analysis. Theory and Applications, pp. 59-71, John Wiley & Sons, Chichester (UK) (2011)
4. Grunsky, E.C.: The interpretation of geochemical survey data. Geochem. Explor. Environ. Anal. **10**, 27–74 (2010)
5. Hennig, C.: fpc: flexible procedures for clustering. R package version 2.1-9 (2014). http://CRAN.R-project.org/package=fpc

6. Palarea-Albaladejo, J., Martín-Fernández, J. A.: zCompositions: Imputation of Zeros and Non-detects in Compositional Data Sets. R package version 1.0.3 (2014). http://CRAN.R-project.org/package=zCompositions
7. R Core Team: R: a language and environment for statistical computing. R Foundation for Statistical Computing, Vienna, Austria (2013). http://www.R-project.org/
8. Templ, M., Hron K., Filzmoser, P. robCompositions: Robust Estimation for Compositional Data. Manual and package, version 1.9.0 (2011). http://cran.r-project.org/web/packages/robCompositions/index.html
9. Templ, M., Hron K., Filzmoser P.: robCompositions: an R-package for robust statistical analysis of compositional data. In: Pawlowsky-Glahn, V., Buccianti, A. (eds.) Compositional Data Analysis. Theory and Applications, pp. 341–355, John Wiley & Sons, Chichester (UK) (2011)

A Compositional Approach to Allele Sharing Analysis

I. Galván-Femenía, J. Graffelman and C. Barceló-i-Vidal

Abstract Relatedness is of great interest in population-based genetic association studies. These studies search for genetic factors related to disease. Many statistical methods used in population-based genetic association studies (such as standard regression models, t-tests, and logistic regression) assume that the observations (individuals) are independent. These techniques can fail if independence is not satisfied. Allele sharing is a powerful data analysis technique for analyzing the degree of dependence in diploid species. Two individuals can share 0, 1, or 2 alleles for any genetic marker. This sharing may be assessed for alleles identical by state (IBS) or identical by descent (IBD). Starting from IBS alleles, it is possible to detect the type of relationship of a pair of individuals by using graphical methods. Typical allele sharing analysis consists of plotting the fraction of loci sharing 2 IBS alleles versus the fraction of sharing 0 IBS alleles. Compositional data analysis can be applied to allele sharing analysis because the proportions of sharing 0, 1 or 2 IBS alleles (denoted by p_0, p_1, and p_2) form a 3-part-composition. This chapter provides a graphical method to detect family relationships by plotting the isometric log-ratio transformation of p_0, p_1, and p_2. On the other hand, the probabilities of sharing 0, 1, or 2 IBD alleles (denoted by k_0, k_1, k_2), which are termed Cotterman's coefficients, depend on the relatedness: monozygotic twins, full-siblings, parent-offspring, avuncular, first cousins, etc. It is possible to infer the type of family relationship of a pair of individuals by using maximum likelihood methods. As a result, the estimated vector $\hat{\boldsymbol{k}} = (\hat{k}_0, \hat{k}_1, \hat{k}_2)$ for each pair of individuals forms a 3-part-composition and can be plotted in a ternary diagram to identify the degree of relatedness. An R package has

I. Galván-Femenía (✉) · C. Barceló-i-Vidal
Department of Computer Science, Applied Mathematics and Statistics,
Universitat de Girona, Campus Montilivi P-IV, 17071 Girona, Spain
e-mail: ivan.galvan@udg.edu

C. Barceló-i-Vidal
e-mail: carles.barcelo@udg.edu

J. Graffelman
Department of Statistics and Operations Research, Universitat Politècnica de Catalunya,
Avinguda Diagonal 647, 6th Floor, 08028 Barcelona, Spain
e-mail: jan.graffelman@upc.edu

J.A. Martín-Fernández and S. Thió-Henestrosa (eds.), *Compositional Data Analysis*, Springer Proceedings in Mathematics & Statistics 187,
DOI 10.1007/978-3-319-44811-4_5

been developed for the study of genetic relatedness based on genetic markers such as microsatellites and single nucleotide polymorphisms from human populations, and is used for the computations and graphics of this contribution.

Keywords Allele sharing · Identical by state · Identical by descent · Cotterman's coefficients · Ternary diagram · Isometric log-ratio transformation

1 Introduction

The application of statistics in genetics and molecular biology has become an active field of research over the last decades. Studies of family relationships in genetic data analysis are crucial in population-based genetic association studies [5]. The main aim of research in these association studies is to find genetic factors related to disease. These studies assume that individuals from human populations are independent. The dependence between individuals, e.g., the presence of related individuals in a database, can invalidate the statistical methods applied in association studies such as regression models or t-tests. Thus, it is important to detect the degree of relatedness of a pair of individuals in the database. This can help to avoid such dependence by removing one individual of the detected pair.

Genetic datasets are composed of genetic markers that are helpful to find the possible DNA regions related to a disease of interest such as cancer. Single nucleotide polymorphisms (SNPs) and microsatellites or short tandem repeats (STRs) are common genetic markers in population-based genetic association studies [9]. SNPs are common throughout the human genome; they occur once every 300 nucleotides on average in the DNA sequences formed by the bases adenine (A), cytosine (C), guanine (G), and thymine (T). SNPs are mostly used for large scale genetic association studies. We give an example of a SNP for three individuals who have the following DNA sequences on a pair of chromosomes at a specific locus: ID1 = (CC**G**ATC, CC**A**ATC), ID2 = (CC**G**ATC, CC**G**ATC), and ID3 = (CC**A**ATC, CC**A**ATC). Note that for the first individual the sequences differ only at the third base pair for the alleles G and A and for the second and the third individual the same alleles appear at the third base. Thus, the third position is a SNP, in this case a **G/A** polymorphism. The three individuals have a SNP coded by the genotypes GA, GG, and AA, respectively, at this specific locus.

On the other hand, microsatellites are short DNA sequences that are repeated. The length of the repeated DNA sequences is constant for each STR and ranges from 2 to 6 nucleotides. Unrelated individuals have genetic variability because their alleles of determined regions of DNA vary. As a result, microsatellites are very powerful to distinguish each individual from the population due to the presence of genetic variability between individuals. Particularly, the number of the repeated sequences across STRs varies between unrelated individuals. For this reason, they are also used for forensic DNA studies. There are two ways to code an STR: by recording the total size in base pairs of the repeating sequences; or by considering only the number of repeats of a particular sequence. For instance, an individual has the following DNA sequences on a pair of chromosomes at a specific locus: ID1 = (ATTATTATTATT, ATTATTATTCCC). This is a trinucleotide repeat ATT that can be coded as an STR of (4, 3) repeats or as an STR of size (12, 9).

Allele sharing analysis is a classical technique for analyzing the degree of dependence between individuals [2]. Two diploid individuals can share 0, 1, or 2 alleles for any genetic marker. The larger the number of shared alleles between a pair of individuals across genetic markers, the more likely they are to be closely related. Thus, individuals from the same family share on average more alleles than unrelated individuals. Allele sharing is called identical by state (IBS) if the DNA composition of the alleles is identical but the alleles do not necessarily come from a common ancestor. It is called identical by descent (IBD) if the alleles originate from a common ancestor. A pair of IBD alleles is necessarily a pair of IBS alleles, but not the reverse. We consider both methods for relatedness research in the rest of this paper.

The estimation of the probabilities of sharing 0, 1, or 2 IBD alleles for a pair of individuals is not trivial. These probabilities depend on the genotypes and the allele frequencies of the population and should be estimated. The maximum likelihood method is considered to be the best procedure for estimating IBD probabilities [17]. Family relationships can be identified by comparing the estimated IBD probabilities with the theoretical Cotterman coefficients (denoted by k_0, k_1, k_2) [3, 15, 16]. We remark that the Cotterman coefficients can be considered in the simplex, S^3, and we show the representation of the estimates of the Cotterman coefficients in a ternary diagram.

This chapter is organized as follows. Section 2 gives an overview of the IBS allele sharing analysis and the application of ilr-coordinates. Section 3 presents the basic principles of the IBD allele sharing and the representation of the family relationships in a ternary diagram. Sections 2 and 3 treat examples of IBD and IBS studies with microsatellite and SNP data. An R package that can simulate data for the studies is discussed (still in development). Finally, Sect. 4 summarizes the principal conclusions of this contribution.

2 Identical by State Studies

IBS studies ignore if the alleles of any pair of individuals are derived from a common ancestor. The IBS sharing of a pair of individuals can be calculated from the genotype data. Then, two individuals share 0 IBS alleles if they have no alleles in common (e.g., AA and GT); share 1 IBS allele if one individual has only a single allele in common with the other individual (e.g., AA and AT or AA and TA; the position of the alleles is irrelevant), and 2 IBS alleles if they have identical genotypes (e.g., AA and AA). Occasionally, the number of shared alleles may be missing (NA) if some individual has missing genotyping information (e.g., AA and NA, or NA and NA).

This approach is usually considered for all the pairs of individuals from a human population across genetic markers. Then, for each pair of individuals we have a vector of 0, 1, or 2 shared alleles as large as the number of genetic markers in the database. Consequently, it is possible to build a vector $\boldsymbol{p}$ of the proportions of shared alleles (0, 1, 2) for each pair of individuals denoted by $\boldsymbol{p} = (p_0, p_1, p_2)$. Classical IBS allele sharing consists of plotting the proportion of sharing 2 IBS alleles (p_2) versus the proportion of sharing 0 IBS alleles (p_0) for all pairs of individuals from a given human population [14]. This graphical method is powerful to detect family relationships by observing the pairs of individuals with higher values of p_2. It is known that family relationships of degree zero, that is, monozygotic twins (MZ), usually have values of p_2 close to 1. Parent-offspring pairs (PO) usually have values of p_0 close to 0. Full-siblings (FS), half-siblings (HS), avuncular (AV), and grandparent–grandchild (GG) are also family relationships that can be detected by this graphical method. Unrelated individuals (UN) usually have higher values of p_0. However, the plot p_2 versus p_0 ignores the constraint $p_0 + p_1 + p_2 = 1$ and the relative information of the component p_1.

For this reason, we propose the isometric log-ratio (ilr) transformation [4] of the vector (p_0, p_1, p_2) in order to preserve the relative information of the 3 parts. The resultant ilr-coordinates can be plotted to detect family relationships in IBS studies.

By construction, the first ilr-coordinate (z_1) interprets the balance between p_2 and p_0. Note that this coordinate captures the information of the graphical method explained before by Rosenberg [14]. The second ilr-coordinate (z_2) corresponds to the balance between p_0, p_2, and p_1. The ilr-coordinates are defined as follows:

$$\begin{cases} z_1 = \frac{1}{\sqrt{2}} \ln\left(\frac{p_2}{p_0}\right) \\ z_2 = \frac{1}{\sqrt{6}} \ln\left(\frac{p_0 p_2}{p_1^2}\right) \end{cases} \tag{1}$$

MZ, PO, FS, HS, AV, GG pairs have higher values of z_1, whereas unrelated individuals have lower values of z_1.

2.1 Example

We present an R package called **IBS.IBD.studies**. This package contains a sample from the Maya population of 25 individuals extracted from a world-wide database from the Noah A. Rosenberg Research lab at Stanford University [13, 14]. This world-wide database is derived from the Human Genome Diversity Cell Line Panel (HGDP) [1]. For each individual, the sample from the Maya population includes 5 columns with their individual code number assigned by the HGDP (ID), the population code number assigned by Rosenberg's lab (Pop.Code), the population name (Pop.Nam), the geographic information (Geographic), and the region of the population (Region). The genetic information consists of 377 microsatellites (STRs) labeled by their respective "locus names" (D12S1638, D14S1007, …). Table 1 shows a glance at the database of the Maya population.

Table 1 A glance at the Maya population

ID	Pop.code	Pop.Nam	Geographic	Region	D12S1638	D14S1007	D9S1779	D9S1825	…
854	86	Maya	Mexico	AMERICA	120	130	142	129	…
854	86	Maya	Mexico	AMERICA	120	124	124	129	…
855	86	Maya	Mexico	AMERICA	126	NA	124	137	…
855	86	Maya	Mexico	AMERICA	120	NA	124	129	…
856	86	Maya	Mexico	AMERICA	126	130	124	135	…
856	86	Maya	Mexico	AMERICA	126	128	124	129	…
857	86	Maya	Mexico	AMERICA	126	128	124	135	…
857	86	Maya	Mexico	AMERICA	126	128	124	129	…
⋮	⋮	⋮	⋮	⋮	⋮	⋮	⋮	⋮	

Each individual from the Maya population is listed in two consecutive lines. For instance, the genotype for the individual ID = 854 for the STR D12S1638 is (120, 120). The allele "120" indicates the total size in base pairs of the repeating DNA sequence. An individual whose genotyping information is missing, is coded by NA (not available).

We use the functions `allelesharing()` and `percentages()` from the **IBS.IBD.studies** R package. The first function computes the shared IBS alleles for each genetic marker and for each pair of individuals. Using `percentages()`, the proportions of sharing 0, 1, or 2 IBS alleles (p_0, p_1, p_2) for each pair of individuals from the sample are obtained.

Figure 1a plots the fraction of loci sharing 2 IBS alleles (p_2) versus the fraction of loci sharing 0 IBS alleles (p_0) for all pairs of individuals from the Maya population. The family relationships documented by Rosenberg [14] are represented in different colors. Observe that outlying individuals correspond to family relationships of the first degree (PO and FS) and the second degree (AV or GG); relationships of the third degree such as first cousins (FC) are more difficult to detect and are mixed with unrelated individuals (UN).

Figure 1b shows the representation in ilr-coordinates of all pairs of individuals of the Maya population. According to the first ilr-coordinate (z_1), the distance between the two PO pairs and the UN pairs from the Maya population equals three units, whereas the distance between the FS pair and the UN pairs is approximately one unit. According to the second ilr-coordinate (z_2), the distance between PO and UN is approximately two units. The z_1 ilr-coordinate facilitates the detection of family relationships which separates related individuals from unrelated individuals. Outlying pairs with large values of the z_1 coordinate usually represent related individuals. Comparing Fig. 1a and b, we note that PO pairs are more outlying in ilr-coordinates.

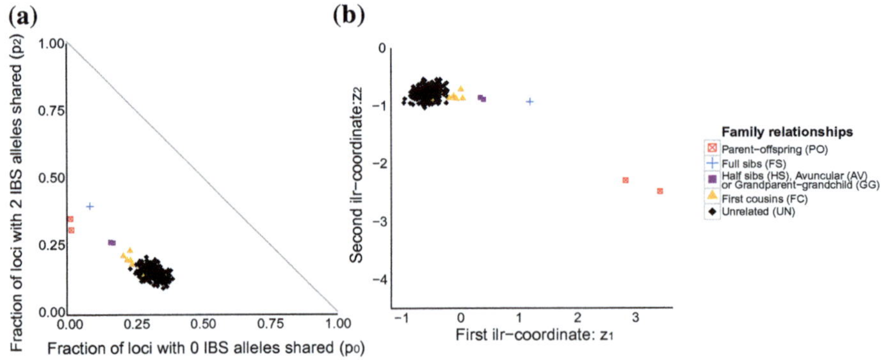

Fig. 1 IBS graphics for all pairs of individuals from the Maya population. **a** Plot of the fraction of loci sharing 2 IBS alleles (p_2) versus the fraction of loci sharing 0 IBS alleles (p_0). **b** ilr-coordinates (z_1, z_2)

3 Identical by Descent Studies

We have shown that IBS studies offer graphical methods to detect relatedness between individuals from human populations. Here, we present IBD studies in order to identify accurately which type of family relationship belongs to each pair of individuals. The degree of relatedness can be inferred by considering the number of IBD alleles shared. The probabilities of sharing 0, 1, or 2 IBD alleles are called Cotterman's coefficients [3] and are denoted by k_0, k_1, and k_2, respectively. These probabilities depend on the relatedness: monozygotic twins (MZ), parent-offspring (PO), full-sibs (FS), half-sibs (HS), avuncular (AV), grandparent-grandchild (GG), first cousins (FC), or unrelated individuals (UN) are presented in Fig. 2 (top). Note that HS, AV, and GG have exactly the same Cotterman coefficients and is not possible to distinguish them, unless we build a pedigree tree for the given human population under study. The set of IBD probabilities has the simplex as its domain. For this reason, it is possible to represent all the family relationships in a ternary diagram as is shown in Fig. 2 (bottom).

In practice, genetic data contains information for estimating the IBD probabilities. However, these probabilities depend on the genotypes and the allele frequencies of

Type of Relative	Degree	k_0	k_1	k_2
Monozygotic twins	0	0	0	1
Parent-offspring	1	0	1	0
Full-siblings	1	1/4	1/2	1/4
Half-siblings/ avuncular/ grandchild-grandparent	2	1/2	1/2	0
First cousins	3	3/4	1/4	0
Unrelated	∞	1	0	0

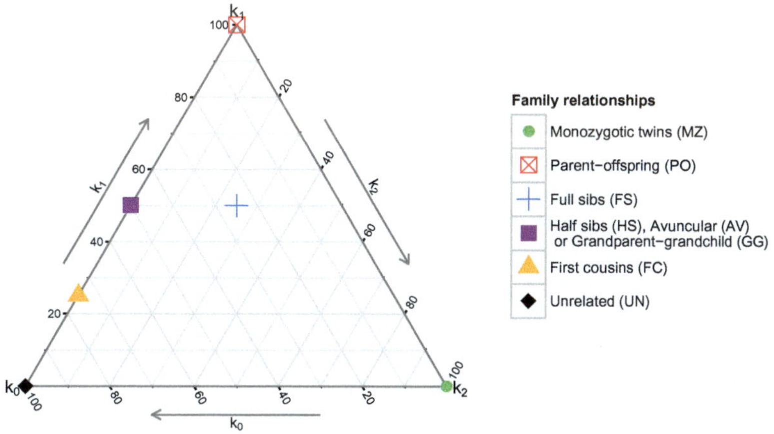

Fig. 2 *Top* Cotterman's coefficients for the different type of family relationship and degree of relatedness. *Bottom* Representation of the Cotterman coefficients in a ternary diagram

the human population under study. For this reason, a good procedure for estimating the Cotterman coefficients is needed. If the estimated IBD probabilities are close or coincide with the theoretical Cotterman coefficients from Fig. 2 (top) for a given relationship, then that relationship is inferred. Many articles have been found in the literature on the estimation of Cotterman's coefficients. Maximum likelihood estimation is the best estimation method [10, 17]. We do not detail the steps of this method in this contribution, as the ML estimation is summarized in Chapter 6 [7]. In this chapter, we compute the maximum likelihood estimates ($\hat{k}_0$, $\hat{k}_1$, and $\hat{k}_2$, respectively) using the function `cotterman()` from the **IBS.IBD.studies** R package. This function uses the optimization routines from the **Rsolnp** R package [6].

Once the estimates of $\hat{k}_0$, $\hat{k}_1$, and $\hat{k}_2$ are obtained, graphical methods such as the plot of $\hat{k}_1$ versus $\hat{k}_0$ [12] or $\hat{k}_2$ versus $\hat{k}_1$ [11] are commonly used to identify the degree of relatedness. These plots separate the related individuals from the unrelated individuals, however, they ignore the relative information of the remaining part of the 3-part-composition $\hat{\boldsymbol{k}} = (\hat{k}_0, \hat{k}_1, \hat{k}_2)$. Therefore, it seems logical to plot the Cotterman coefficients in a ternary diagram as an additional graphical method in order to identify relatedness. The ternary diagram has the advantage of showing all Cotterman coefficients simultaneously, in contrast to the $\hat{k}_1$ versus $\hat{k}_0$ or $\hat{k}_2$ versus $\hat{k}_1$ plots that represent only two of them. This way, the information of the IBD probabilities is preserved for all pairs of individuals and the family relationships are inferred by comparing the estimates with the theoretical values of k_0, k_1 and k_2 plotted in Fig. 2 (bottom). The Cotterman coefficients can be plotted in a ternary diagram by using the function `ggplot()` from the **ggtern** R package [8] as shown by the examples below.

3.1 Examples

In this section, we use simulated and empirical datasets and plot them in a ternary diagram in order to identify relationships. The functions `simSNP()` and `children()` from the R package **IBS.IBD.studies** can be used to simulate genetic marker data with given family relationships. First, we generate a sample of 20 unrelated individuals with 1000 genetic markers. The function `simSNP()` simulates random single nucleotide polymorphisms (SNPs), giving categorical variables with the three genotypes AA, AB, and BB. All SNPs have a minor allele frequency of 0.5. SNPs are simulated independently under the assumption of Hardy–Weinberg equilibrium: $p_A^2 + 2p_Ap_B + p_B^2 = 1$ [5]. Thus, each SNP is a random sample of a multinomial distribution of size 20 (the number of unrelated individuals). The theoretical genotype probabilities of AA, AB, and BB are 0.25, 0.5, and 0.25, respectively. Each individual is labeled by 'ID' as shown in Table 2.

Once the sample is generated, we use the function `children()` to build a pedigree tree as follows. Because we know that a child always has received one allele from the father and one allele from the mother, the function `children()`

Table 2 A glance at the simulated dataset

Individuals	SNP1	SNP2	SNP3	SNP4	...	SNP1000
ID1	AA	AB	AB	AB	...	BB
ID2	AA	BB	AB	BB	...	AB
ID3	BB	AB	AB	AA	...	BB
ID4	AB	BB	AB	AA	...	AB
ID5	AB	AB	BB	AA	...	BB
⋮	⋮	⋮	⋮	⋮	⋮	⋮
ID20	BB	AA	AB	AA	...	BB

chooses one allele randomly across SNPs from the individual 'ID1' and one allele from the individual 'ID2' to generate a child. This new individual is labeled by 'ID21A'. Analogously, we produce another child labeled by 'ID21B'. Thus, 'ID21A' and 'ID21B' are a full-siblings pair. We complete the pedigree tree by generating a child (labeled by 'ID22') of the individuals 'ID2' and 'ID3' in order to originate two half-siblings pairs, which correspond to the pairs 'ID21A'–'ID22' and 'ID21B'–'ID22'. An additional family relationship was created, a duplicated individual of 'ID15' (labeled by 'ID23'); hence, 'ID15' and 'ID23' represent a monozygotic twin pair. Thus, the new simulated population consists of 24 individuals. The simulated relationships are composed of one MZ pair, six PO pairs, one FS pair and two HS pairs as shown in Fig. 3.

We again study the Maya population (Sect. 2.1) as an empirical example.

Figure 4 shows the ternary diagrams of the estimated Cotterman coefficients $(\hat{k}_0, \hat{k}_1, \hat{k}_2)$ for all pairs of individuals of the simulated human population and the Maya population respectively. All the family relationships of degree zero, one, and two are close to the theoretical probabilities described in Fig. 2. However, relationships of degree three such as FC in Fig. 4 (right) are difficult to discriminate from UN or AV. The ternary diagram of the Maya data reveals higher values for k_1 and lower values for k_0 of the FS and HS pairs in comparison with the simulated data. We suggest this is due to the fact that the simulated data consists mainly of entirely

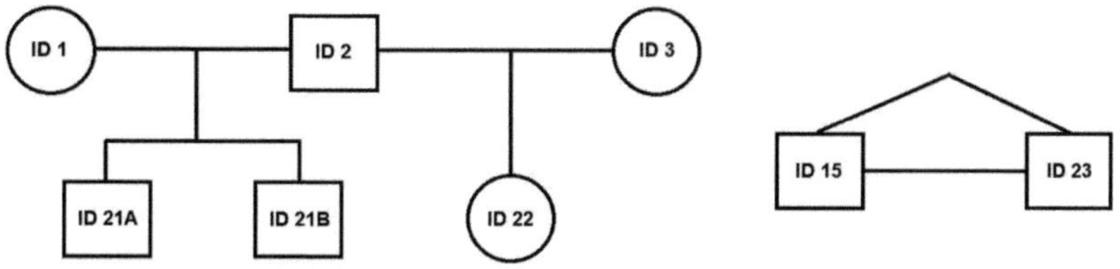

Fig. 3 The simulated family relationships. *Left* a pedigree tree consisting of six PO pairs, one FS pair and two HS pairs. *Right* a MZ pair

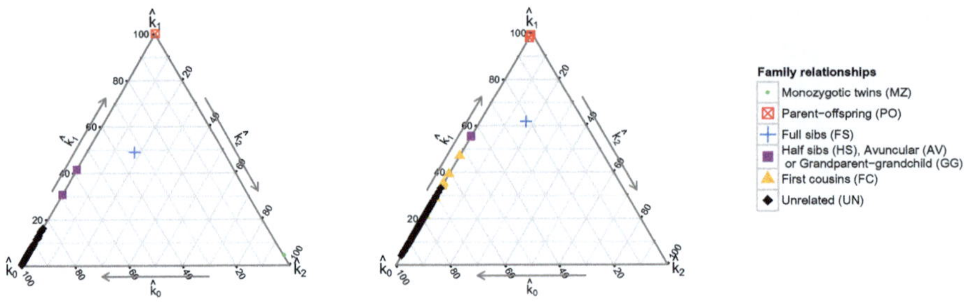

Fig. 4 Ternary diagrams of the estimated Cotterman coefficients $(\hat{k}_0, \hat{k}_1, \hat{k}_2)$ for all pairs of individuals from the simulated data (*left*) and the Maya population (*right*)

unrelated individuals, whereas the Maya individuals derive from common ancestors, and are probably inbred to a certain extent.

4 Conclusion

This chapter is focused on the annotation of the relatedness between individuals in genetic data. We stress the importance of the detection of family relationships as a tool for quality control in population-based genetic association studies. The statistical methods used in these studies can fail if the dependence between individuals is not documented. We have shown two classical approaches in relatedness research and we have applied tools from compositional data for the identification of family relationships.

First, the IBS allele sharing analysis provides a graphical tool for detecting related individuals. This graph plots the proportion of sharing 2 IBS alleles versus the proportion of sharing 0 IBS alleles. We propose an additional graphical method by using the isometric log-ratio transformation of the vector of proportions of sharing 0, 1, or 2 IBS alleles. We plot the ilr-coordinates for all pairs of individuals and use this plot to detect relationships.

Finally, the IBD allele sharing analysis offers an accurate estimation of relatedness by using Cotterman's coefficients. Plots of $\hat{k}_1$ versus $\hat{k}_0$ or $\hat{k}_2$ versus $\hat{k}_1$ are used to represent graphically the family relationships from a population. These plots ignore the constraint $\hat{k}_0 + \hat{k}_1 + \hat{k}_2 = 1$ and we state that ternary diagrams may be useful to identify family relationships. The theoretical values of k_0, k_1 and k_2 form reference points in the ternary diagram for the standard relationships.

Acknowledgments We thank the referees and the editors for their comments on the manuscript. This study was supported by grant CODARSS MTM2012-33236 (2013–2015) of the Spanish Ministry of Education and Science.

References

1. Cavalli-Sforza, L.L.: The human genome diversity project: past, present and future. Nature Rev. Genet. **6**, 333–340 (2005)
2. Chakraborty, R., Jin, L.: Determination of relatedness between individuals using DNA fingerprinting. Hum. Biol. **65**(6), 875–895 (1993)
3. Cotterman, C.W.: Relative and human genetic analysis. Sci. Mon. **53**, 227–234 (1941)
4. Egozcue, J.J., Pawlowsky-Glahn, V., Mateu-Figueras, G., Barceló-Vidal, C.: Isometric logratio transformations for compositional data analysis. Math. Geol. **35**(3), 279–300 (2003)
5. Foulkes, A.S.: Applied Statistical Genetics with R. Springer (2009)
6. Ghalanos, A., Theussl, S.: Rsolnp: general non-linear optimization using augmented Lagrange multiplier method. R package version **1**, 15 (2014)
7. Graffelman, J., Galván-Femenía, I.: An application of the isometric log-ratio transformation in relatedness research. In: Martín-Fernández J, A., Thió-Henestrosa, S. (eds.) Compositional Data Analysis, Springer Proceedings in Mathematics & Statistics **187**, (2016)
8. Hamilton, N.: ggtern: An Extension to 'ggplot2', for the Creation of Ternary Diagrams. R package version 1.0.6.0 (2015). http://CRAN.R-project.org/package=ggtern
9. Laird, N.M., Lange, C.: The fundamentals of modern statistical genetics. Springer (2011)
10. Milligan, B.G.: Maximum-likelihood estimation of relatedness. Genetics **163**, 1153–67 (2003)
11. Moltke, I., Albrechtsen, A.: RelateAdmix: a software tool for estimating relatedness between admixed individuals. Bioinformatics **30**, 1027–8 (2014)
12. Nembot-Simo, A., Graham, J., McNeney, B.: CrypticIBDcheck: an R package for checking cryptic relatedness in nominally unrelated individuals. Source Code Biol. Med. **8**, 5 (2013)
13. Rosenberg, N.A.: Rosenberg lab at Stanford University (2002). http://www.stanford.edu/group/rosenberglab/diversity.html
14. Rosenberg, N.A.: Standardized subsets of the HGDP-CEPH human genome diversity cell line panel, accounting for atypical and duplicated samples and pairs of close relatives. Ann. Hum. Genet. **70**, 841–847 (2006)
15. Thompson, E.A.: Estimation of pairwise relationships. Ann. Hum. Genet. **39**, 173–188 (1975)
16. Thompson, E.A.: Estimation of relationships from genetic data. In: Rao, C.R., Chakraborty, R. (eds.) Handbook of Statistics, vol. 8, pp. 255–269. Elsevier Science, Amsterdam (1991)
17. Weir, B.S., Anderson, A.D., Hepler, A.B.: Genetic relatedness analysis: modern data and new challenges. Nature Rev. Genet. **7**, 771–780 (2006)

References

An Application of the Isometric Log-Ratio Transformation in Relatedness Research

J. Graffelman and I. Galván-Femenía

Abstract Genetic marker data contains information on the degree of relatedness of a pair of individuals. Relatedness investigations are usually based on the extent to which alleles of a pair of individuals match over a set of markers for which their genotype has been determined. A distinction is usually drawn between alleles that are identical by state (IBS) and alleles that are identical by descent (IBD). Since any pair of individuals can only share 0, 1, or 2 alleles IBS or IBD for any marker, 3-way compositions can be computed that consist of the fractions of markers sharing 0, 1, or 2 alleles IBS (or IBD) for each pair. For any given standard relationship (e.g., parent–offspring, sister–brother, etc.) the probabilities k_0, k_1 and k_2 of sharing 0, 1 or 2 IBD alleles are easily deduced and are usually referred to as Cotterman's coefficients. Marker data can be used to estimate these coefficients by maximum likelihood. This maximization problem has the 2-simplex as its domain. If there is no inbreeding, then the maximum must occur in a subset of the 2-simplex. The maximization problem is then subject to an additional nonlinear constraint ($k_1^2 \geq 4k_0k_2$). Special optimization routines are needed that do respect all constraints of the problem. A reparametrization of the likelihood in terms of isometric log-ratio (ilr) coordinates greatly simplifies the maximization problem. In isometric log-ratio coordinates the domain turns out to be rectangular, and maximization can be carried out by standard general-purpose maximization routines. We illustrate this point with some examples using data from the HapMap project.

Keywords Genetic marker · Identity-by-state · Identity-by-descent · Hardy–Weinberg equilibrium · Composition · Closure · Ternary plot · Isometric log-ratio transformation

---

J. Graffelman (✉)
Department of Statistics and Operations Research, Universitat Politècnica de Catalunya, Avinguda Diagonal 647, 6th floor, 08028 Barcelona, Spain
e-mail: jan.graffelman@upc.edu

I. Galván-Femenía
Department of Computer Science, Applied Mathematics and Statistics, Universitat de Girona, Campus Montilivi P-IV, 17071 Girona, Spain
e-mail: ivan.galvan@udg.edu

J.A. Martín-Fernández and S. Thió-Henestrosa (eds.), *Compositional Data Analysis*, Springer Proceedings in Mathematics & Statistics 187,
DOI 10.1007/978-3-319-44811-4_6

1 Introduction

Methods for investigating the degree of relatedness of a pair of individuals form an active area of research in statistical genetics. Relatedness studies are based on the idea that genetic markers carry information regarding the family relationships of the individuals involved. There are several reasons for performing a relatedness study. First of all, such studies serve to verify documented relationships between individuals. If sufficient genetic information is available, a relatedness study may reveal that a pair of putative sibs is in fact a pair of half-sibs or a twin pair. An important criterion in relatedness studies is the degree to which a pair of individuals shares alleles over a set of genetic markers. If two individuals share many alleles at many loci then it becomes more likely that they are closely related. In the most extreme case, if all alleles at all loci coincide for a pair of individuals, then the pair is, supposing sufficiently polymorphic loci, in theory a monozygotic twin pair. However, some caution is called for, because 100 % coincidence will also arise if two registers in the database are accidentally duplicated, or if a biological sample has been genotyped twice in the laboratory.

Secondly, relatedness is a reason of concern in gene-disease association studies for statistical reasons. The presence of related individuals violates the independence assumption that underlies many statistical techniques used in these studies, such as chi-square tests on contingency tables, logistic regression, and others. It is thus of interest to investigate the possible relatedness of the individuals in the sample prior to applying tests for association. If relatedness is detected, one individual of each related pair may be removed in order to maintain independence.

A distinction can be drawn between relatedness studies that are based on alleles that are identical by state (IBS) and alleles that are identical by descent (IBD). Two alleles are IBS if they are the same irrespective of their provenance, a situation that is statistically often referred to as a "match." In this contribution we restrict ourselves to IBD allele-sharing. Two alleles are IBD if they are IBS and have descended from the same parent. Because a child receives one allele from each parent, it shares one allele IBD with its father and one allele IBD with its mother. It is not possible for the child to receive zero or two IBD alleles from the father. The probabilities of sharing 0, 1, or 2 IBD alleles for a given relationship are called *Cotterman coefficients*, and denoted by k_0, k_1 and k_2, respectively. If X denotes the number of IBD alleles for a parent–offspring (PO) pair then its probability distribution is $k_0 = P(X = 0) = 0$, $k_1 = P(X = 1) = 1$ and $k_2 = P(X = 2) = 0$. For other relationships it is only slightly more involved to obtain the theoretical IBD probabilities. Let α/β represent the paternal alleles and A/B the maternal alleles of a couple sharing two alleles IBS. This couple can have four possible types of children (α/A), (α/B), (β/A), and (β/B), and the number of IBD alleles for each possible pair of sibs is shown in Table 1.

From this table it is easily inferred that the IBD sharing probabilities for a pair of full sibs are given by $k_0 = \frac{1}{4}$, $k_1 = \frac{1}{2}$ and $k_2 = \frac{1}{4}$. IBD probabilities can be estimated from the genotype data by maximum likelihood, as described in Sect. 2. If the esti-

Table 1 Number of IBD alleles for all possible pairs of sibs descendant from a $(\alpha/\beta,\ A/B)$ couple

	α/A	α/B	β/A	β/B
α/A	2	1	1	0
α/B	1	2	0	1
β/A	1	0	2	1
β/B	0	1	1	2

mated probabilities coincide with or are close to a set of theoretically known IBD probabilities for a given relationship, then that relationship is inferred.

The structure of the remainder of this chapter is as follows. First we review the maximum likelihood estimation of IBD probabilities. Then we present a reparametrization of the likelihood in terms of isometric log-ratios. We give some detailed examples with data from the HapMap project. Finally we discuss our results and provide some references.

2 Maximum Likelihood Estimation of IBD Probabilities

Good accounts of the ML estimation of IBD probabilities are given by Thompson [11] and Weir et al. [12]. We briefly review ML estimation here in order to provide a self-contained chapter. The Cotterman coefficients can be obtained in a similar way as outlined in the introduction for all standard family relationships and are given in Table 2.

Let G_1 and G_2 be the pair of genotypes observed at a locus for two individuals, and let q (0, 1 or 2) represent the number of IBD alleles. By the law of total probability

Table 2 IBD probabilities or Cotterman coefficients for some standard family relationships (MZ = Monozygotic twins, PO = Parent–offspring, FS = Full sibs, HS = Half-sibs, AV = Avuncular, GG = Grandparent-grandchild, FC = First cousins, UN = Unrelated)

Relationship	k_0	k_1	k_2
MZ	0	0	1
PO	0	1	0
FS	$\frac{1}{4}$	$\frac{1}{2}$	$\frac{1}{4}$
HS	$\frac{1}{2}$	$\frac{1}{2}$	0
AV	$\frac{1}{2}$	$\frac{1}{2}$	0
GG	$\frac{1}{2}$	$\frac{1}{2}$	0
FC	$\frac{3}{4}$	$\frac{1}{4}$	0
UN	1	0	0

we have

$$P(G_1 \cap G_2 | k_0, k_1, k_2) = P(G_1 \cap G_2 | q = 0)\, k_0 + P(G_1 \cap G_2 | q = 1)\, k_1 + P(G_1 \cap G_2 | q = 2)\, k_2. \tag{1}$$

The probabilities $P(G_1 \cap G_2 | q = 0)$ depend on the genotypes of the individuals and are calculated from the allele frequencies in the population. We denote different alleles by the letters i, j, l and m and let p_i, p_j, p_l and p_m be their corresponding allele frequencies. We denote genotypes by i/j, the slash separating the alleles found on the homologous chromosomes. For example, if $G_1 = i/i$ and $G_2 = i/i$, then under the assumption of Hardy–Weinberg equilibrium (with probabilities p_i^2 and $2p_ip_j$ for homozygous and heterozygous genotypes respectively) we obtain

$$P(G_1 = i/i \cap G_2 = i/i | q = 0) = P(G_1 = i/i)\, P(G_2 = i/i) = p_i^2 p_i^2 = p_i^4,$$
$$P(G_1 = i/i \cap G_2 = i/i | q = 1) = P(G_1 = i/i)\, P(G_2 = i/i | G_1 = i/i | q = 1) = p_i^2 p_i = p_i^3,$$
$$P(G_1 = i/i \cap G_2 = i/i | q = 2) = P(G_1 = i/i) = P(G_2 = i/i) = p_i^2.$$

These probabilities are also determined for all other genotype pairs $((i/i, i/j)$, $(i/i, j/j)$, etc.) and the results are given in Table 3. If there are n independent genetic markers, then the likelihood function for a pair of individuals can be written as

$$L(k_0, k_1, k_2 | G_1 \cap G_2) = \prod_{i=1}^{n} (d_{0i} k_0 + d_{1i} k_1 + d_{2i} k_2), \tag{2}$$

where the coefficients d_{0i}, d_{1i} and d_{2i} depend on the nature of the pair (possibilities given in Table 3) and on the allele frequencies of the corresponding markers. For example, if for one marker both individuals are homozygous (an $(i/i, i/i)$ pair) then the contribution to the likelihood function is $p_i^4 k_0 + p_i^3 k_1 + p_i^2 k_2$, with p_i the ith allele frequency of that marker. Taking logarithms, we search to maximize the log-likelihood function

Table 3 Probabilities of observing 0, 1 or 2 IBD alleles for all possible genotype pairs

Pair	Shared alleles	$q = 0$	$q = 1$	$q = 2$
$(i/i, i/i)$	2	p_i^4	p_i^3	p_i^2
$(i/i, j/j)$	0	$p_i^2 p_j^2$	0	0
$(i/i, i/j)$	1	$2p_i^3 p_j$	$p_i^2 p_j$	0
$(i/i, j/m)$	0	$2p_i^2 p_j p_m$	0	0
$(i/j, i/j)$	2	$4p_i^2 p_j^2$	$p_i p_j (p_i + p_j)$	$2p_i p_j$
$(i/j, i/m)$	1	$4p_i^2 p_j p_m$	$p_i p_j p_m$	0
$(i/j, m/l)$	0	$4p_i p_j p_m p_l$	0	0

$$l(k_0, k_1, k_2 | G_1 \cap G_2) = \sum_{i=1}^{n} \ln \left(d_{0i} k_0 + d_{1i} k_1 + d_{2i} k_2 \right), \tag{3}$$

where k_0, k_1 and k_2 are the parameters to be estimated, and the coefficients d_{0i}, d_{1i}, d_{2i} are obtained by substituting the sample allele frequencies in accordance with the type of pair.

3 Reparametrization of the Likelihood in Isometric Log-Ratios

The maximization of the likelihood in Eq. (3) is not trivial, as the following constraints need to be taken into account: $0 \leq k_i \leq 1$, $\sum_{i=0}^{2} k_i = 1$, and $k_1^2 \geq 4k_0k_2$. The last inequality follows from the assumption of absence of inbreeding [10]. This is a maximization problem that has an arrow-headed subset of the simplex as its domain, as is shown by the gray region in Fig. 1a. Standard R functions like `optim` or `nlminb` from the **stats** package [7] assume the domain of the objective function to be rectangular. For these R functions the simplex constraint $k_0 + k_1 + k_2 = 1$ is a problem. The range for k_i is not simply the [0, 1] interval but the limits depend on the values of the other parameters. Besides this linear constraint, at the same time the nonlinear inequality $k_1^2 \geq 4k_0k_2$ must also be taken into account. If the standard functions are used for the problem at hand, then the algorithm will typically step outside the feasible region leading to numerical errors. The function `solnp` from the R-package Rsolnp [4] solves general nonlinear programming problems and allows for inequalities and nonlinear equalities, and can handle our maximization problem. Figure 1 also represents the standard relationships whose compositions are given in Table 2. All these relationships are at the edge of the feasible region.

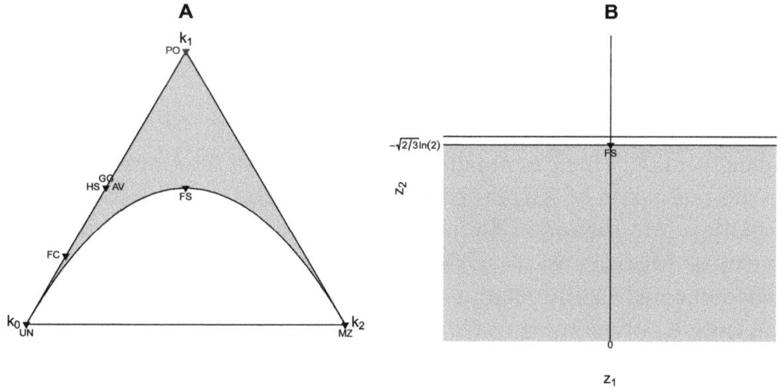

Fig. 1 Domain of the likelihood function (*gray*) in a ternary plot representation (**a**) and in isometric log-ratio coordinates (**b**)

The parabola delimiting the feasible region described by $k_1^2 = 4k_0k_2$ can be recognized as coinciding with the second dimension of the isometric log-ratio coordinates [1] of composition $\mathbf{k} = (k_0, k_1, k_2)$. The same parabola is also of relevance in studies of Hardy–Weinberg equilibrium [5], but with a different genetic interpretation of the composition. This suggests that we might reparameterize the likelihood in terms of the isometric log-ratio coordinates in order to obtain a rectangular domain for the likelihood. This simplifies the maximization problem, as it can now be solved using R's general-purpose optimization routines `optim` and `nlminb`.

We use the isometric log-ratio transformation, calculating the coordinates as follows:

$$z_1 = \frac{1}{\sqrt{2}} \ln\left(\frac{k_0}{k_2}\right), \quad z_2 = \frac{1}{\sqrt{6}} \ln\left(\frac{k_0 k_2}{k_1^2}\right). \tag{4}$$

The inverse relationships are given by

$$(k_0, k_1, k_2) = \mathscr{C}(e^{\sqrt{2}z_1}, e^{\frac{1}{2}\sqrt{2}z_1 - \frac{1}{2}\sqrt{6}z_2}, 1), \tag{5}$$

where $\mathscr{C}$ is the closure operator. This gives the reparameterized log-likelihood

$$l(z_1, z_2 | G_1 \cap G_2) = \sum_{i=1}^{n} \left(\ln\left(d_{0i} e^{\sqrt{2}z_1} + d_{1i} e^{\frac{1}{2}\sqrt{2}z_1 - \frac{1}{2}\sqrt{6}z_2} + d_{2i}\right) - \ln\left(1 + e^{\sqrt{2}z_1} + e^{\frac{1}{2}\sqrt{2}z_1 - \frac{1}{2}\sqrt{6}z_2}\right)\right). \tag{6}$$

The nonlinear constraint $k_1^2 \geq 4k_0k_2$ becomes a linear inequality for the second ilr-coordinate

$$z_2 \leq -\sqrt{\frac{2}{3}} \ln(2). \tag{7}$$

The domain of the reparameterized log-likelihood is shown in Fig. 1b.

4 Examples

We use data from the Mexican population of phase III of the HapMap project [9] to illustrate the estimation of IBD probabilities in log-ratio coordinates. The data consist of genotype information of 86 individuals (mostly parent–offspring trios) from Los Angeles of Mexican ancestry. Several scholars have analyzed these data, and many undocumented family relationships have been reported [3, 6, 8]. We filtered single nucleotide polymorphisms (SNPs) from the genome-wide HapMap database as follows. SNPs significant in a chi-square test for Hardy–Weinberg equilibrium ($\alpha = 0.05$) were excluded to avoid possible genotyping error. SNPs with a minor allele frequency below 0.4 were also excluded in order to guarantee a set of sufficiently polymorphic markers. We sampled 5,000 SNPs at random from this sub-

Table 4 ML estimation of IBD probabilities of a FS pair, using 5,000 SNPs, with initial point (0.575, 0.400, 0.025). Iteration histories with the log-likelihood (l) for maximization in original and in log-ratio coordinates by `solnp` and `nlminb` respectively

`solnp`				
It.	l	$\hat{k}_0$	$\hat{k}_1$	$\hat{k}_2$
1	−9483.1290	0.41422	0.48104	0.10474
2	−9368.1777	0.18452	0.56753	0.24796
3	−9366.4621	0.21746	0.52776	0.25478
4	−9366.4615	0.21697	0.52798	0.25505
5	−9366.4615	0.21697	0.52798	0.25505
`nlminb`				
It.	l	$\hat{z}_1$	$\hat{z}_2$	
0	−9671.5480	2.217130	−0.983749	
1	−9503.2604	1.528570	−1.708930	
2	−9446.4083	0.538662	−1.567190	
3	−9415.8636	0.214144	−1.361630	
4	−9367.0124	−0.184897	−0.705093	
5	−9366.5961	−0.108992	−0.696269	
6	−9366.4802	−0.127274	−0.667618	
7	−9366.4659	−0.113317	−0.666588	
8	−9366.4622	−0.116864	−0.661281	
9	−9366.4617	−0.114033	−0.661312	
10	−9366.4615	−0.114846	−0.660250	
11	−9366.4615	−0.114261	−0.660282	
12	−9366.4615	−0.114416	−0.660107	
13	−9366.4615	−0.114323	−0.660105	
14	−9366.4615	−0.114323	−0.660105	

set, and consider the resulting dataset as a set of approximately independent and highly polymorphic markers. We consider two pairs chosen from this database as an example.

The first pair is a presumably unrelated pair of individuals with identifiers NA19662 and NA19685. This pair was inferred to be a FS pair [6] using the program RELPAIR [2]. Here we reanalyze the relationship of this pair using ML estimation of the IBD probabilities, with initial point (0.575, 0.400, 0.025). The iteration history and log-likelihood (l) are given in Table 4 for the maximization of Eq. (3) with `solnp` and for the maximization in ilr-coordinates (Eq. (6)) with `nlminb`. Both algorithms converge to the same maximum. Back-transformation of the final ilr-coordinates (−0.1143, −0.6601) gives the same estimates $\hat{k}_0 = 0.217$, $\hat{k}_1 = 0.528$ and $\hat{k}_2 = 0.255$, which confirms the hypothesis of a FS pair. The level curves of the log-likelihood function are shown in Fig. 2a and show the maximum as an interior point.

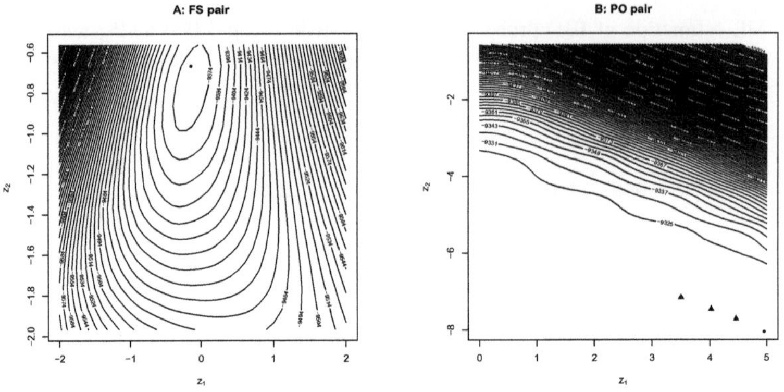

Fig. 2 Level curves of the log-likelihood function in ilr-coordinates for a FS pair and a PO pair

We consider a second example of another undocumented relationship between NA19660 and NA19685 that was inferred to be a parent–offspring (PO) pair [6]. This example differs from the previous one because now the reference relationship (PO with $k_0 = 0$, $k_1 = 1$ and $k_2 = 0$) is outside the simplex. Results for the maximization in original and log-ratio coordinates are shown in Table 5.

Function `solnp` gives estimates $\hat{k}_0 = 0.0018$, $\hat{k}_1 = 0.9982$ and $\hat{k}_2 = 0.0000$ which suggests a PO pair. The maximization in log-ratio coordinates shows that

Table 5 ML estimation of IBD probabilities of a PO pair, using 5,000 SNPs, with initial point (0.575, 0.400, 0.025). Iteration histories for maximization of the likelihood function (l) in original and in log-ratio coordinates by `solnp` and `nlminb` respectively

`solnp`				
It.	l	$\hat{k}_0$	$\hat{k}_1$	$\hat{k}_2$
1	−9489.1607	0.29473	0.70527	0.00000
2	−9320.4669	0.00184	0.99816	0.00000
3	−9320.4669	0.00184	0.99816	0.00000
`nlminb`				
It.	l	$\hat{z}_1$	$\hat{z}_2$	
0	−9680.5924	2.21713	−0.98375	
1	−9324.1179	0.24058	−3.86971	
2	−9322.1662	1.56428	−4.90716	
3	−9321.5402	2.13094	−5.41660	
4	−9320.7389	3.37902	−6.55765	
5	−9320.5206	4.20767	−7.31837	
6	−9320.4712	4.73722	−7.80468	
7	−9320.4672	4.91901	−7.97160	
8	−9320.4671	4.94894	−7.99905	
9	−9320.4671	4.95042	−8.00037	

z_1 increases and z_2 decreases until the change in the log-likelihood drops below the tolerance used. Back-transformation of the coordinates gives the result $\hat{k}_0 = 0.0018$, $\hat{k}_1 = 0.9982$ and $\hat{k}_2 = 0.0000$ which coincides with the estimates obtained by `solnp`. Figure 2b shows the level curves of the log-likelihood function together with the maximum found (marked with a dot). As z_2 decreases at some point the log-likelihood function becomes very flat.

For the FS pair, different initial points were used and all converged to the same maximum. For the PO pair, different starting points often give different solutions in ilr-coordinates. The maxima found according to the iteration histories in Tables 4 and 5 are marked by a point. For the PO pair, the solutions obtained from three different additional initial points are shown by black triangles. In ilr-coordinates these solutions differ, but all correspond to an area where the log-likelihood function is very flat. Back-transformation of all solutions gives however, up to five decimals, the same IBD probabilities.

5 Discussion

In this contribution we have shown that isometric log-ratio coordinates can be used to simplify a genetic maximization problem that has the simplex or a subset of the simplex as its domain. In statistics there are more likelihood functions that are subject to a unit sum constraint on the parameters, and that may possibly be simpler to maximize in log-ratio coordinates, such as likelihoods based on the multinomial distribution. Maximization of likelihoods in coordinates may therefore have a wider applicability then suggested by the specific genetic problem dealt with here. The main advantage is that the irregular domains of the likelihood function in the simplex can become rectangular when expressed in ilr-coordinates. Most standard maximization functions can then handle the maximization problem, whereas specialized software is needed for maximizing the likelihood function over a subset of the simplex.

We note that the theoretical IBD probabilities for the most common family relationships fall on the edge of the simplex (see Fig. 1a). Only the FS relationship is an interior point of the simplex and all other relationships from Table 2 are outside the simplex because zeros are not admitted. Thus, when maximizing in coordinates, these theoretical probabilities can never be attained. In practice the log-ratio coordinates tend to extreme values for these relationships, and convergence can be slow because the log-likelihood function flattens. Setting an adequate tolerance criterion may help to speed up the convergence.

The ilr-coordinates of the numerical solution found for relationships on the edge of the simplex can vary considerably (see Fig. 2b). However, when back-transformed to IBD probabilities the corresponding relationships can be inferred. In this respect, we note that large negative values for z_1 combined with z_2 at its maximum of $-\sqrt{2/3}\ln(2)$ point to a MZ pair, and large positive values for z_1 combined with z_2 at its maximum of $-\sqrt{2/3}\ln(2)$ suggest an unrelated pair (UN). Avuncular pairs (AV, aunt–niece, etc.), grandparent–grandchild pairs (GG), and half-sibs (HS) can not be distinguished because they all have the same Cotterman coefficients. With the isometric log-ratio transformation proposed here, these relationships are characterized by a positive z_1 and proportionality between the log-ratio coordinates ($z_1 = -\sqrt{3}z_2$).

Acknowledgments This study was supported by grant CODARSS MTM2012-33236 of the Spanish Ministry of Education and Science. We thank Josep Daunis-i-Estadella and an anonymous referee for their comments that helped us to improve the chapter.

References

1. Egozcue, J.J., Pawlowsky-Glahn, V., Mateu-Figueras, G., Barceló-Vidal, C.: Isometric logratio transformations for compositional data analysis. Math. Geol. **35**(3), 279–300 (2003)
2. Epstein, M., Duren, W., Boehnke, M.: Improved inference of relationship for pairs of individuals. Am. J. Hum. Genet. **67**, 1219–1231 (2000)
3. Gazal, S., Sahbatou, M., Perdry, H., Letort, S., Gnin, E., Leutenegger, A.L.: Inbreeding coefficient estimation with dense snp data: comparison of strategies and application to hapmap III. Hum. Hered. **77**(1–4), 49–62 (2014)
4. Ghalanos, A., Theussl, S.: Rsolnp: general non-linear optimization using augmented lagrange multiplier method. R package version 1.15 (2014). http://CRAN.R-project.org/package=Rsolnp
5. Graffelman, J., Egozcue, J.J.: Hardy-weinberg equilibrium: a nonparametric compositional approach. In: Pawlowsky-Glahn, V., Buccianti, A. (eds.) Compositional Data Analysis: Theory and Applications, pp. 208–217. John Wiley & Sons, Ltd. (2011)
6. Pemberton, T.J., Wang, C., Li, J.Z., Rosenberg, N.A.: Inference of unexpected genetic relatedness among individuals in hapmap phase III. Am. J. Hum. Genet. **87**, 457–464 (2010)
7. R Core Team: R: A Language and Environment for Statistical Computing. R Foundation for Statistical Computing, Vienna, Austria (2014). http://www.R-project.org/
8. Stevens, E.L., Baugher, J.D., Shirley, M.D., Frelin, L.P., Pevsner, J.: Unexpected relationships and inbreeding in hapmap phase III populations. PLoS ONE **7**(11) (2012). doi:10.1371/journal.pone.0049575
9. The International HapMap Consortium: Integrating common and rare genetic variation in diverse human populations. Nature **467**, 52–58 (2010)
10. Thompson, E.A.: The estimation of pairwise relationships. Ann. Hum. Genet. **39**, 173–188 (1975)
11. Thompson, E.A.: Estimation of relationships from genetic data. In: Rao, C.R., Chakraborty, R. (eds.) Handb. Stat., vol. 8, pp. 255–269. Elsevier Science, Amsterdam (1991)
12. Weir, B.S., Anderson, A.D., Hepler, A.B.: Genetic relatedness analysis: modern data and new challenges. Nature Rev. Genet. **7**(10), 771–780 (2006)

Recognizing and Validating Structural Processes in Geochemical Data: Examples from a Diamondiferous Kimberlite and a Regional Lake Sediment Geochemical Survey

E.C. Grunsky and B.A. Kjarsgaard

Abstract Geochemical data are compositional in nature and are subject to the problems typically associated with data that are restricted in real positive number space, the simplex. Geochemistry is a proxy for mineralogy, and minerals are comprised of atomically ordered structures that define the placement and abundance of elements in the mineral lattice structure. The arrangement of elements within one or more minerals that comprise rocks, soils, and surficial sediments define a linear model in the Euclidean geometry of real space in terms of their geochemical expression. When methods such as principal component analysis are applied to multielement geochemical data, the dominant components generally reflect features related to mineralogy and describe geologic processes that are both independent and partially codependent. The dominant principal components can be used as a filter to eliminate noise or under-sampled processes in the data. These dominant components can be used to create predictive geological maps, or maps displaying recognizable geochemical processes. Using these techniques, we demonstrate that stoichiometrically controlled geochemical processes can be "discovered" and "validated" from two sets of data, one derived from drill-hole lithogeochemistry of a series of kimberlite eruptions and a second from a suite of granitic, metamorphic, volcanic, and sedimentary rocks.

Keywords Geochemistry · Classification · Lithologic prediction · Compositional data analysis

E.C. Grunsky (✉)
Department of Earth and Environmental Sciences, University of Waterloo, Waterloo N2L 3G1, Canada
e-mail: egrunsky@gmail.com

B.A. Kjarsgaard
Geological Survey of Canada, Ottawa K1A 0E8, Canada

J.A. Martín-Fernández and S. Thió-Henestrosa (eds.), *Compositional Data Analysis*, Springer Proceedings in Mathematics & Statistics 187,
DOI 10.1007/978-3-319-44811-4_7

1 Introduction

The compositional nature of geochemical data requires care when considering relationships between the elemental, oxide, or molecular constituents that define a composition. The use of ratios are essential when making comparisons between elements in systems such as igneous fractionation [25, 32, 35] and the use of log ratios is essential when measuring moments such as variance/covariance in examination of data derived from geochemical surveys [1, 4, 7, 22].

The relationships between the elements of geochemical data are governed by "natural laws" [2] and specifically by stoichiometry and thereby imposing structure within the data. Grunsky and Bacon-Shone [12] have shown that geochemical patterns and trends are closely related to the stoichiometric constraints of minerals.

To effectively interpret geochemical data, a two-phase approach is suggested; that of process discovery, followed by process validation. This tactic identifies geochemical/geological processes that exist in the data, but are not obvious unless robust statistical methods are utilized. The process discovery phase is most effective when carried out using a multivariate approach. Linear combinations of elements related by stoichiometry are generally expressed as strong patterns, whilst random patterns and under-sampled processes show weak or uninterpretable patterns. Grunsky et al. [13] initially demonstrated these concepts using multielement lake sediment geochemical data from the Melville Peninsula area, Nunavut, Canada.

Two examples are presented. The first demonstrates the usefulness of compositional data analysis combined with multivariate statistical analysis for process discovery and validation from drill core geochemistry from the Star kimberlite in Saskatchewan, Canada [11]. Distinct geochemical kimberlite phases can be statistically identified using this approach and lead to efficiencies in the economic evaluation of kimberlite for diamonds.

The second example is the evaluation of lake sediment geochemistry obtained from a regional geochemical survey [19]. The results presented here are from a campaign to reanalyze sample pulps using modern analytical methods including inductively coupled plasma mass spectrometry (ICP-MS). In cases where elements have been analyzed using two or more methods, the elements were evaluated in terms of detection limit suitability and visual examination of the correlation of the element with each method. The elements B, Ge, In, Pd, Re, Ta, and Te were dropped due to large numbers of observations that were reported at less than the lower limit of detection (lld); furthermore, these elements are not key elements observed in typical rock forming minerals.

One of the primary purposes of geochemical data analysis is the recognition of geochemical/geological processes. Processes are recognized by a continuum of variable responses and the relative increase/decrease of these variables. The presence of censored data (values $<$ lld) can, in some cases, affects the results of a process recognition investigation. In the compositional data analysis framework, early on Aitchison [1] recognized the problem of censored data. Martín-Fernández et al. [17, 18] and Hron et al. [16] discuss various replacement options based on the nature of the censored data. Recognizing the difference between missing (i.e., no data) values and censored (<lld) data is crucial in deciding how a replacement value, if any, should be estimated. Details on the methods used are described below.

2 Methods

The methods used in this study were applied in the R programming environment [31] and are documented below. Venables and Ripley [37] provide many useful details and scripts for analyzing, visualizing, and statistically studying data.

2.1 *Process Discovery*

Process discovery involves the use of unsupervised multivariate methods such as principal component analysis (PCA), multidimensional scaling and Random Forests, to name a few. Process discovery with geochemical data reflects the recognition of linear models that reflect the stoichiometry of rock forming minerals and subsequent processes that modify mineral structures (hydrothermal fluids, weathering, groundwater). Additional processes such as fluid dynamics effects can effectively sort minerals according to the energy of the environment and mineral density. The chemistry of minerals is governed by stoichiometry and the relationships are described within the simplex. Geoscientists have long recognized that many geochemical processes can be clearly described using cation ratios that reflect the stoichiometric balances of minerals during formation (e.g., Pearce [25]).

Geochemical data, expressed in elemental form is a proxy for mineralogy. If the mineralogy of a geochemical data set is known, then the proportions of these elements can be determined using linear methods. Grunsky [10] has reviewed some of the normative mineral calculation procedures that are available. An advantage of estimating normative mineral compositions is that the linear combinations of normative minerals will increase the signal to noise ratio in the normative mineral compositions. However, in many cases the compositions of minerals that have a continuum of element substitutions requires assumptions to be made about such compositions, and the resulting estimates may not reflect the actual mineral compositions or abundances. Thus, in this study, we have chosen to use only the geochemical data and use the observed linear patterns as proxies for mineralogy.

Many geochemical datasets contain values that are reported at less than the lower limit of detection and these values are generally termed "censored." The estimation of statistical parameters can be severely affected by censored data and it is useful to find a replacement value that does not bias the estimate of the statistical moments. The R-package "zCompositions" with the function (lrEM) [20, 21] was used to determine suitable replacement values for several of the elements.

When mixtures of minerals (rocks, soils, glacial till, stream sediments) are analyzed for their geochemistry there may be multiple linear processes embedded in the resulting geochemical analyses. Techniques such as principal component analysis can be effectively used to identify these processes. A method of PCA used in this study is a combined R-mode/Q-mode PCA as documented in Grunsky [8].

An essential part of the process discovery phase is a suitable choice of coordinates to overcome the problem of closure. The logcentered transform [1] is a suitable transform for the evaluation of geochemical data. The resulting principal components are orthonormal and reflect linear processes related to stoichiometry. The components are ideal for subsequent process validation. Enhancements to the visualization of groups of data on scatterplots were made using the R library "cluster" and the function "ellipsoidhull," which creates a convex hull around specified groups of data.

2.2 *Process Validation*

Process validation is the methodology employed to verify that a geochemical composition (response) is associated with identified processes. The processes can be in the form of, e.g., lithology, soil character, ecosystem properties, climate, or deeply buried tectonic assemblages. Validation can be in the form of an estimate of likelihood that a composition can be assigned membership to one of the identified processes. This is typically done through the assignment of a class identifier or a measure of probability. Assignment of class membership can be done through the application of techniques such discriminant analysis, logistic regression, neural networks, or Random Forests. There are many other methods available.

An essential part of process validation is the selection of variables that enable efficient classification, which involves selecting variables that maximize the differences between the different classes and minimizes the amount of overlap due to noise or unresolved processes in the data. Within the context of compositional data, variables that are selected for classification require transformation to log-ratio coordinates. The additive log ratio (alr) or the isometric log ratio (ilr) are equally effective for the implementation of classification procedures. The logcentered (clr) transform is not suitable because the covariance matrix of these coordinates is singular. However, analysis of variance (ANOVA) applied to logcentered data enables recognition of the compositional variables (elements) that are most effective for distinguishing between the classes. Choosing an effective alr transform (choice of suitable divisor) or balances for the ilr transform is not a trivial task. Moreover, ANOVA applied to the principal components derived from the logcentered transform can be highly effective at discriminating between the different classes. Subsequent classification can be carried out with far fewer variables based on principal components derived from the logcentered transform.

Initially, an ANOVA (R function "aov") was applied to the logcentered elements or the principal components derived from the logcentered data. In this study, the principal components were used in the analysis of variance.

Classification results can be conveniently expressed as direct class assignment or posterior probabilities in the form of forced class allocation, or as class typicality. Forced class allocation assigns a posterior probability based on the shortest Mahalanobis distance of a compositional observation from the compositional centroid of each class. Class typicality measures the Mahalanobis distance from each class and assigns a posterior probability based on the F-distribution. This approach can result with an observation having a zero posterior probability for all classes, indicating that the composition is not close to the compositions defined by the class compositional centroids. The results shown for this study used the R function, "lda", from which the posterior probabilities were estimated.

The results of cross-validation linear discriminant analysis (LDA) are posterior probabilities of class membership, which are also compositions [1, 23, 34]. As a result, the subsequent derivation of maps displaying posterior probabilities requires a suitable log-ratio transformation to preserve nonnegativity and overcome the constant sum constraint. In this study, the posterior probabilities were transformed using an additive (alr) log-ratio transformation. Ordinary cokriging, using the R **gstat** package [26], was applied to the transformed posterior probabilities followed by a back-transformation for geospatial rendering. However, the alr transform cannot be used to estimate kriging variance [1, 34]. Kriging variance can be estimated by the calculation of the expected value and error variance covariance matrix by Gauss–Hermite integration [24] after which a backtransform can be applied. In this study, kriging variance was not considered and cokriging of the alr-transformed posterior probabilities was applied directly.

Classification accuracies can be assessed through the generation of tables that show the accuracy and errors measured from the estimated classes against the initial classes in the training sets used for the classification.

2.3 Geospatial Coherence

The results from the classification of materials gathered from a geochemical survey should bear a geospatial resemblance to the area sampled. The creation of maps are part of the process validation procedure. If a geospatial rendering of a posterior probability shows no spatial coherence, then it is likely that the classification is difficult to interpret within a geologic context. The most effective way to test this is through the generation and modeling of semivariograms that describe the spatial continuity of a specific class based on the posterior probabilities. If meaningful semivariograms can be created, then geospatial maps of the posterior probabilities can be generated through interpolation using the kriging process. Maps of posterior probabilities may show low overall values but still be spatially coherent. This is also reflected in the classification accuracy matrix that indicates the extent of classification overlap between classes. Geospatial analysis methodology described by Bivand et al. [3] and the gstat package [26] were used to generate the geostatistical parameters and images of the principal components and posterior probabilities from kriging.

3 Examples

3.1 Process Discovery and Validation of Diamondiferous Kimberlites

Grunsky and Kjarsgaard [11] evaluated diamond drill core lithogeochemistry from the Star kimberlite located in Saskatchewan, Canada. The Star kimberlite is a series of kimberlite eruptions with five recognized phases (from oldest to youngest: early Joli Fou, mid-Joli Fou, late Joli Fou, Pense, Cantuar). The early Joli Fou phase contains more macro-diamonds than the other phases, thus making it highly desirable to recognize this phase in the diamond exploration and evaluation process. Figure 1 shows a scatter plot matrix of elements that are typically associated with kimberlite magma, and fractionation processes. The data are presented in cation percent and are untransformed. The scatter plots reveal two distinct patterns. There is a linear pattern of relative enrichment/depletion for all five phases. Additionally, there is a distinct pattern of relative enrichment/depletion associated with the early, mid- and late Joli Fou kimberlites. A separate trend is noted for the Pense and Cantuar phases. The late Joli Fou and the Cantuar kimberlites show the greatest relative enrichment of the four elements shown in Fig. 1. These linear patterns reveal the control that stoichiometry has over the formation of minerals and the associated changes in mineralogy during kimberlite magma contamination (mantle and crustal), and fractionation processes, as described below. Such patterns are well documented by Pearce [25].

The lithogeochemical data were also evaluated in a multivariate process discovery context by the application of a principal component analysis (PCA) on logcentered data. The eigenvalues of the PCA indicated that the first three components accounted for 77 % of the overall data variation, with the first two accounting for 66 %. Figure 2 shows the PCA biplot for the data in which there are three notable compositional trends. The first is a trend toward the positive PC1 and negative PC2 axis in which there is a relative enrichment of P, Nb, La, Th, and Zr that represents kimberlite magma and fractionation processes. Mineral phases associated with this trend include apatite and perovskite. The Pense and Cantuar phases appear to have the highest kimberlite magmatic component, and exhibit more extensive fractionation than the mid- and late Joli Fou phases. The trend of the mid- and late Joli Fou phases toward the positive PC1 and PC2 axes represents relative enrichment in K, Rb, and Na. This represents crustal contamination through the assimilation of feldspar minerals that exist in the upper crust through which the kimberlite magma ascended; alkali feldspars do not crystallize from kimberlite magma. The third trend along the negative PC1 axis shows relative enrichment in Si, Ni, Mg, Cr, and Co and represents primary olivine and Cr-spinel in the kimberlite, but importantly contamination of the kimberlite magma by the Earth's lithospheric mantle. These elements are typically associated with the mantle minerals olivine, orthopyroxene, Cr-diopside, chromite, and Cr-pyrope garnet. The principal component biplot of the first two elements reveals distinctive information about the processes that have affected the kimberlite geochemical compositions.

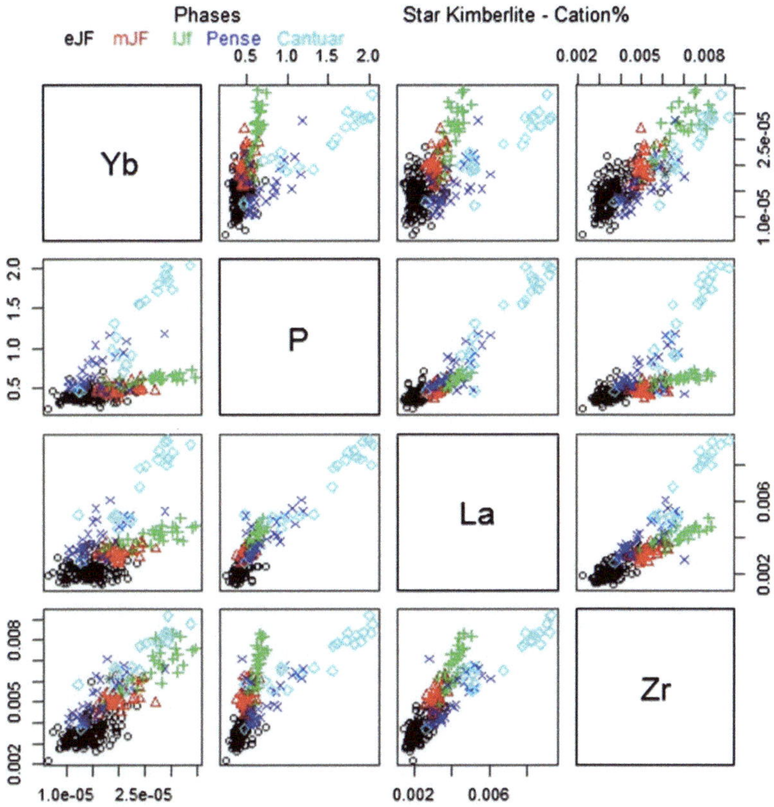

Fig. 1 Scatter plot matrix of elements that are typically associated with kimberlite magma fractionation from the Star kimberlite geochemistry. Note the linear relationships between these elements. There is one linear trend associated with the evolution of the eJF-mJF-lJF eruptions and a second linear trend associated with the Pense and Cantuar eruptions. Originally published in Grunsky and Kjarsgaard [11]. © crown Copyright 2008. Published with kind permission of Elsevier. All Rights Reserved.

A linear discriminant analysis was carried out using an additive log ratio based on Ga as the divisor. Log-ratio theory demonstrates that the choice of any divisor will yield the same results when using classification methods [1]. In this study, Ga was chosen because it appears to be neutral with respect to the processes observed in the PC1-PC2 biplot, i.e., it plots near the origin on the PC1-PC2 plot. Furthermore, ANOVA of individual elements in a logcentered transform indicate Ga (and Yb) has the least discriminating ability between the five phases [11], making it highly suitable as the divisor. This is also consistent with the idea of using a conserved element to model igneous processes [25, 32]. Table 1 shows the results of the classification using the first three principal components derived from the alr transformed data using Ga as the divisor. The overall accuracy is over 91 % with minimal overlap. This is shown

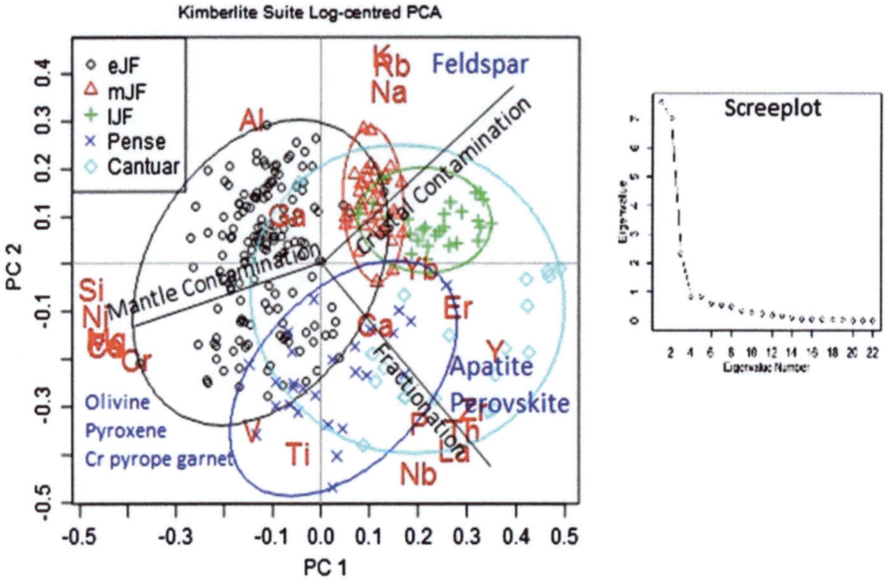

Fig. 2 PCA biplot of PC1-PC2 for the logcentered Star kimberlite geochemical data. Three distinct trends are displayed: kimberlite fractionation, crustal contamination, and mantle contamination. The screeplot shows that most of the data variation is contained in the first two components. The five phases are each enclosed by an elliptical hull to indicate distinctiveness, and overlap

Table 1 Classification accuracy of the Star kimberlite phases based on linear discriminant analysis of the first seven principal components

Count accuracy					
	Cantuar	eJF	lJF	mJF	Pense
Cantuar	20	0	0	0	2
eJF	0	146	0	4	4
lJF	0	0	24	4	0
mJF	0	2	1	37	0
Pense	1	4	0	0	22
% accuracy					
Cantuar	90.91	0.00	0.00	0.00	9.09
eJF	0.00	94.81	0.00	2.60	2.60
lJF	0.00	0.00	85.71	14.29	0.00
mJF	0.00	5.00	2.50	92.50	0.00
Pense	3.70	14.81	0.00	0.00	81.48
Overall accuracy (%)	91.88				

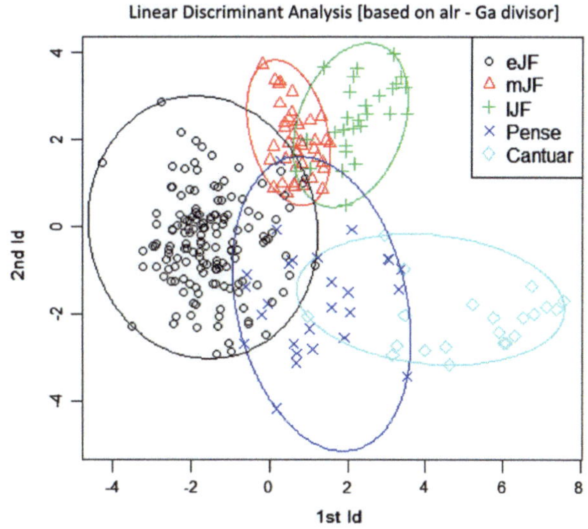

Fig. 3 Plot of linear discriminant function scores 1 and 2 for the Star kimberlite geochemical data. The five phases are each enclosed by an elliptical hull to indicate distinctiveness, and overlap. This is summarized in Table 1

graphically in Fig. 3 where the linear discriminant scores are plotted on the first two discriminant axes.

The effectiveness of process recognition using principal component analysis enables the identification of distinct groups of observations associated with different processes. In this example, the relative proportion of kimberlite magma, and crustal and mantle contamination components are identified, in addition to magma fractionation processes. Furthermore, these processes are clearly identified with specific element associations, which in turn are characteristic of the mineral(s) that are associated with a specific process.

3.2 Predictive Mapping Regional Geology Using Lake Sediment Geochemistry

3.2.1 Process Discovery

Reanalysis of lake sediments was carried out over three 1:250,000 scale map areas (NTS 65A, 65B, 65C) in southern Nunavut Territory, Canada [19]. The geology of two of the NTS sheets (65A, 65B) had been mapped at 1:250K by Eade [5, 6] and is shown in Fig. 4 along with the lake sediment sampling sites. NTS sheet 65C was compiled at a regional 1:500K scale [33] from 1960s reconnaissance one million scale maps by Tella et al. [33] although they did not distinguish between two important granitic intrusion types; the Hudson granite (1.83 Ga) and the Nueltin granite (1.75 Ga) suites as identified and characterized by Peterson et al. [27, 28].

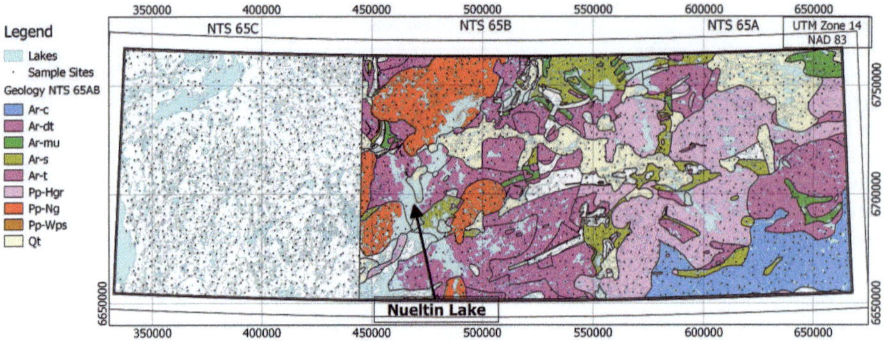

Fig. 4 Geology and lake sediment sample sites for NTS 65A/B/C. The geology of NTS65C has not been mapped in enough detail to distinguish the two Proterozoic granitoid suites, namely the Nueltin granite (*orange* Pp-Ng) and Hudson granite (*light purple* Pp-Hgr)

The three map sheets are located within the southern Hearne Province, a poorly understood terrane located between the central Hearne supracrustal terrane (to the north), which is dominated by ca. 2.7–2.65 Ga mafic-to-felsic, oceanic volcanic rocks, and younger tonalite to granite plutons, and the Trans-Hudson orogen (to the south), which forms the northern boundary of the Superior Province. The southern Hearne domain is dominantly comprised of Archean tonalitic and charnokitic gneisses, approximately 2.8 Ga in age. However, strong evidence for fragments of much older crust, up to 3.3 Ga, has been found in the form of inherited Archean zircons and Sm–Nd model ages obtained from Proterozoic post-orogenic plutons of the Hudson granite, intruded at about 1.83 Ga. Nueltin rapakivi granite (ca. 1.75 Ga) is also present in the area. Previously, van Breemen et al. [36] distinguished Proterozoic plutons in this area on the basis of archived hand samples, field descriptions, and reconnaissance geochronological data. West of Nueltin Lake (NTS 65C), however, identification of individual plutons was hampered by poor exposure and a lower frequency of archival material.

In this study, the lake sediment geochemistry from two of the map sheets (NTS 65A, 65B) were tagged based on the known bedrock lithology derived from the detailed geology by Eade ([5, 6]; Table 2) of the sample site for the purpose of predicting the geology of NTS65C. The lake sediment sample sites from NTS 65C were tagged as "unknown." A suite of 45 elements (Ag, Al, As, Au, Ba, Be, Bi, Ca, Cd, Ce, Co, Cr, Cs, Cu, Fe, Ga, Hf, Hg, K, La, Li, Mg, Mn, Mo, Na, Nb, Ni, P, Pb, Rb, S, Sb, Sc, Se, Sn, Sr, Th, Ti, Tl, U, V, W, Y, Zn, Zr) were determined by inductively coupled plasma emission spectroscopy/mass spectrometry (ICP-ES/MS) after an aqua regia partial digestion [19]. The mineralogy of these rocks are dominantly quartz (Si), feldspars (K, Na, Ca, Al, Si), micas (muscovite, biotite) that contain varying amounts of K and Fe and ferromagnesian minerals (e.g., pyroxene, olivine, hornblende,) whose compositions are controlled by varying amounts of Fe, Mg, Ca, Mn, Ti, Cr. The two significant granitoid suites of the area are the Nueltin and Hudson granites [29]. The Nueltin suite of intrusions range from alkali granite, syenogranite and monzogranite,

Table 2 Dominant lithologies used for study in NTS 65A/B

Legend	#Sites	Description
Ar_mu	50	Archean mafic rocks
Ar_c	131	Archean carbonate rocks
Ar_dt	266	Archean diorite
Ar_t	221	Archean tonalite
Ar_s	111	Archean sediments
Pp_Hgr	266	Paleoproterozoic Hudson granite
Pp_Hu	49	Paleoproterozoic Hurwitz sediments
Pp_Nq	178	Paleoproterozoic Nueltin Granite
Qt	189	Quaternary sediments
Pp-Wps	27	Paleoproterozoic psammitic gneiss
Unknown	833	Sample sites from NTS 65C

and are comprised of quartz, alkali feldspar with strongly enrichment in Zr–Y–U–Th and rare earth elements (REE: La–Ce–Pr–Nd–Sm–Eu–Gd–Tb–Dy–Ho–Er–Tm–Yb–Lu). The granitoid rocks representing the Hudson granite are of similar major element composition and although there is compositional overlap between the two granitoid suites, the Hudson granite contains lesser amounts of heavy rare earth elements (HREE), notably Tb, Dy, Ho, Er, Tm Yb, Lu, Y, which from a mineralogical perspective suggests lower modal xenotime and/or allanite contents as compared to Nueltin granitoids.

Initially, the data were transformed using the centered logratio [1] from which a principal component analysis was carried out as part of the "discovery process" approach. A tabular summary of results documents that the first seven components account for 72.8 % of the overall variation in the data (Appendix A). This is also illustrated in Fig. 5, where a "screeplot" of the ordered eigenvalues are shown. The first seven components display a steep decay indicating that these components account for most of the variability of the data. Appendix A also shows R-score values for each element over each principal component. The magnitude and sign of the R-scores indicate the relative significance of a given element with respect to each other for a given component. The significance of the R-scores is directly associated with the magnitude of each eigenvalue. The relative contributions shown in Appendix A indicate the relative significance of an element over the principal components and the absolute contributions indicate the relative significance of an element within a given principal component. Furthermore, Appendix A details that many elements with high loadings are observed in the first five components. However, some elements such as Au, Bi, Sb, and S have a significant amount of variability accounted for in lesser components (PC5, 7, 9, 10) which likely represent significant, but under-sampled processes. These components could be highly useful vectors for Au mineral exploration follow-up. The content of Appendix A, for the first three principal components is graphically expressed in Fig. 6a, b.

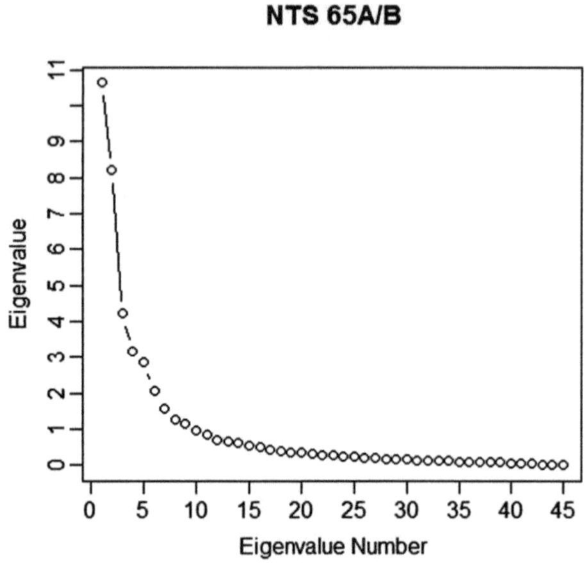

Fig. 5 Ordered eigenvalue plot ("screeplot") of from the principal component analysis of the lake sediment geochemistry sample suite

Figure 6a shows a principal component biplot of PC1-PC2. The R-scores for each element are located throughout the plot and correspond to the scores shown in Appendix A. The scores of the individual observations (sample sites) are plotted with a symbol and color that indicates the lithology at the sample site. The legend displays the coding for the sample scores. The two lithologies of interest in this study, the Hudson granite (Pp-Hgr) and Nueltin granite (Pp-Ng), are demarcated by ellipses that describe convex hulls for these bedrock geology units. The sample sites from NTS 65C are shown as gray dots throughout the figure. Importantly, many sample sites tagged as Nueltin granite (Pp-Ng) occur in the negative PC1—negative PC2 quadrant and show relative enrichment in Ag–Zn–Mo–Y–La–Ce–Be–P–U. For greater clarity, Fig. 6b, c show biplots of the PC scores for the lake sediment sites coded as Hudson and Nueltin granites, respectively. The relative positions of these scores with respect to the principal component loadings of the elements shows that the sample sites associated with the Hudson granite (Fig. 6b) show a contrast in relative enrichment of K–Ga–Rb–Li–Cs–Sn–Th and Ca–Sr–Sb–Hg–S–Ba–Cu. The latter group of elements likely reflects the influence of supracrustal rocks in the lake sediment composition. Figure 6c shows the PC scores of the lake sediment sample sites coded with the Nueltin granite. The trend of the elliptical hull that encompasses the Nueltin observations in Fig. 6c is approximately orthogonal to the Hudson elliptical hull in Fig. 6b. The relative enrichment trend and contrast for the Nueltin granite sites are Ag–Zn–Mo–Y–La–Ce–Be in contrast with Mg–Cr–Sc–Ti–As–Nb–Th–Co–Au. The biplot of PC2-PC3 displayed in Fig. 6d shows a distinct trend of relative enrichment of U–La–Ce–Sn–Th–Rb, which is associated with the Nueltin granite. In summary, the lake sediment data shows a strong, 1 to 1 correspondence between

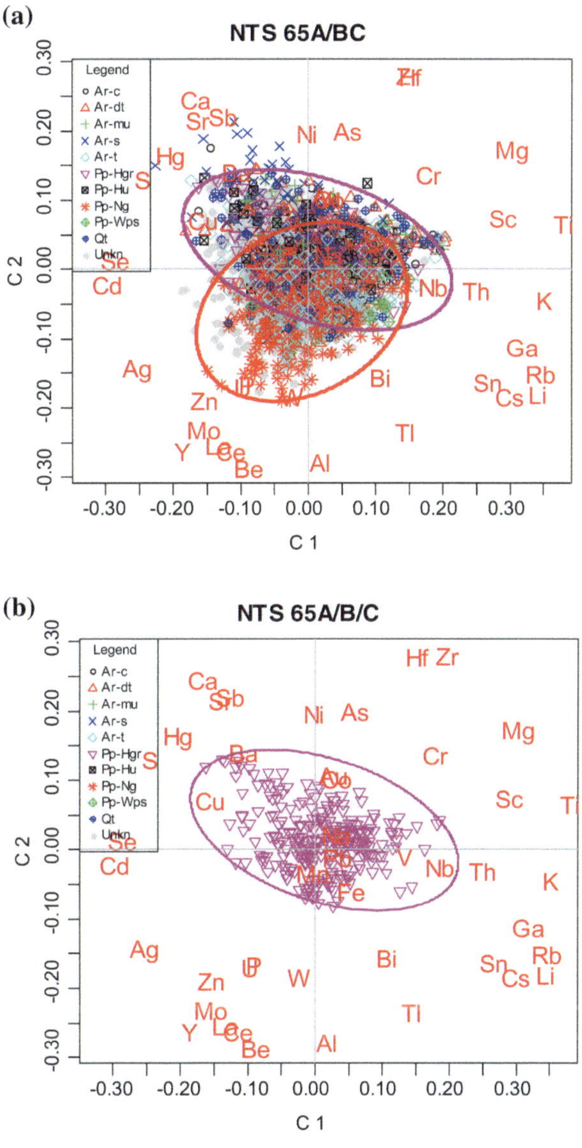

Fig. 6 **a** Principal component biplot of PC1-PC2 for the lake sediment geochemical data. Elliptical hulls are shown for the Nueltin granite (Pp-Ng) [*red*] and the Hudson granite (Pp-Hgr) [*magenta*]. **b** Principal component biplot of PC1-PC2 for the lake sediment geochemical data. Only the Hudson granite (Pp-Hgr) scores are shown in this biplot. **c** Principal component biplot of PC1-PC2 for the lake sediment geochemical data. Only the Nueltin granite (Pp-Ng) scores are shown in this biplot. **d** Principal component biplot of PC2-PC3 for the lake sediment geochemical data. Elliptical hulls for the Nueltin granite (Pp-Ng) and the Hudson granite (Pp-Hgr) are shown in this biplot. Many of the "unknown" sample sites show relative uranium enrichment within the *ellipse* that defines the range of compositions associated with the Nueltin granite

the Nueltin granite and positive lake sediment anomalies in specific incompatible elements, notably REE and Y, in the well-characterized area around Nueltin Lake.

Sample sites tagged with the Hudson granite lithology (Pp-Hgr) occur near the center of the biplot. The ellipsoid hull that surrounds these points displays a trend from the negative PC1–positive PC2 quadrant (Ba–Cu–Ca signature) to the positive PC1–negative PC2 quadrant (Nb–Th–K–Ga–Sn–Cs–Li–Rb signature). The association of Archean sediments (Ar-s) in the negative PC1–positive PC2 quadrant of the biplot

Fig. 6 (continued)

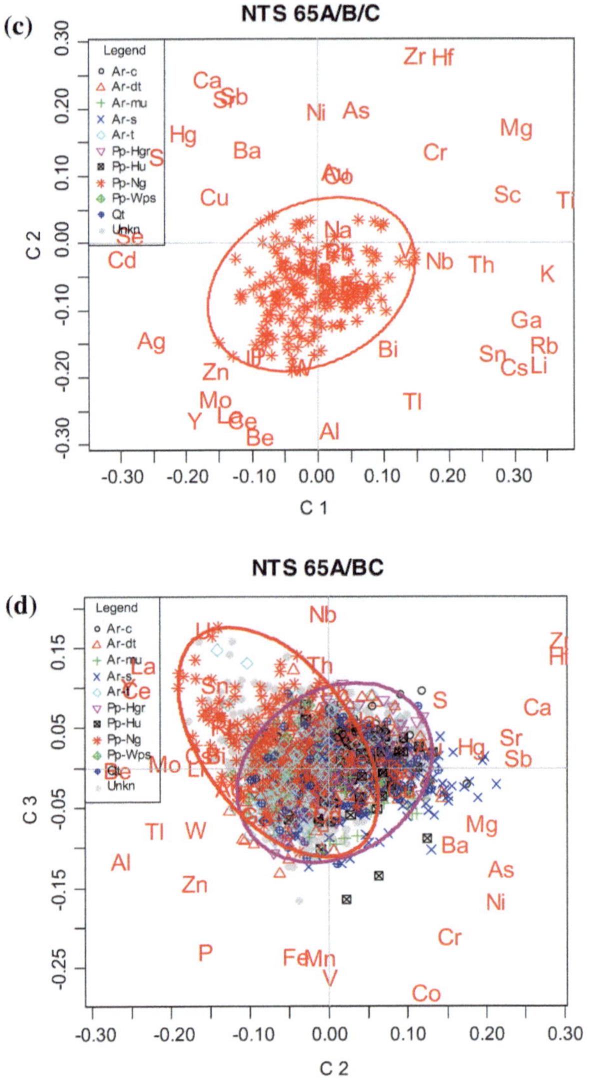

may be indicative of some mixing of the bedrock lithologies in the lake sediments, although this association may also be the result of contamination or partial melting of Archean sediment. The biplots also show that there is significant overlap of principal component scores for most of the sample sites over a wide range of lithologies. Thus, it is difficult to define distinct element associations and trends for many of these lithologies, using the methodologies outlined above.

Following procedures outlined by Grunsky et al. [15], a thorough geostatistical analysis of the principal component scores was carried out using the R-package, "gstat." Modeled semivariograms and variogram maps (semivariograms created with an azimuthal range of 0–359° at 1° increments) were examined from which kriged images of the principal components were created. Figure 7a shows a variogram map for the first principal component. The image shows a moderate anisotropy with a maximum sill at 75° East of North and a minimum sill at 145° East of North. A semivariogram was modeled at 75° east of north and shown in Fig. 7b. This model was used to generate the kriged image shown in Fig. 7c, which was integrated into the Quantum GIS package [30]. High values of PC1 represent relative enrichment

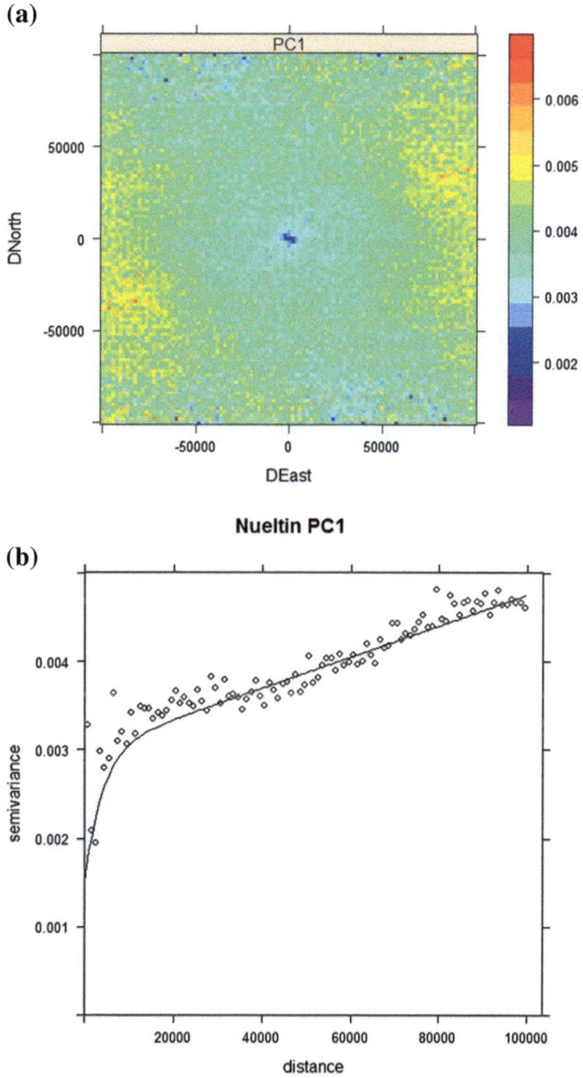

Fig. 7 **a** Variogram map of the first principal component derived from the lake sediment geochemical data. The map demonstrates anisotropy trending in a south-easterly direction. **b** Modeled semivariogram of the first principal component derived from the lake sediment geochemical data. The semivariogram is derived from an anisotropic search direction with a combined short-range and long-range model. **c** Interpolated image of the first principal component using the modeled semivariogram as illustrated in Fig. 7b. Grid is meters. The range of PC scores cover most of the dominant rock types and the variability appears to be independent of underlying lithologies

(c)

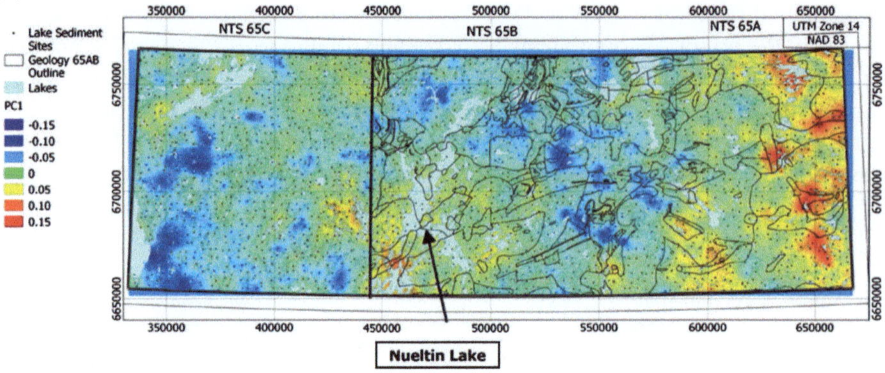

Fig. 7 (continued)

of K, Ti, Ga, Rb, Cs, Li, Sn, Sc, Mg and Th. This relative enrichment overlies Quaternary sediments (Qt), Archean tonalite (Ar-t) and Archean carbonate (Ar-c) rocks. Negative PC1 scores represent relative enrichment of Cd, Se, S, Hg, Ag and overlie a range of Archean lithologies including tonalite and diorite (Ar-t, Ar-dt) and Quaternary sediments (Qt). The lack of direct correspondence of the first principal component with any specific lithology is not unexpected as it represents the dominant geochemical variability that is typical of the major lithologies throughout the area.

The variogram map of the second principal component (Fig. 8a) shows anisotropy with a minimum sill trending at 170° and a maximum sill trending at 45°. A modeled semivariogram (Fig. 8b) modeled at 45° east of north was used to create the kriged image of Fig. 8c. The positive values of PC2 are associated with Archean sediments and carbonate (Ar-s, Ar-c) and the negative values are associated with the Nueltin granite (Pp-Ng) as is shown in Fig. 6a. Note the excellent correspondence between the negative PC2 scores around Nueltin Lake, and the spatial distribution of Nueltin granite as shown in Fig. 4.

Although not shown, the third principal component has positive scores associated with Archean diorite (Ar-dt) and Nueltin granite (Pp-Ng), and the negative scores have a distinct association with the Hurwitz Group sediments (Pp-Hu). Positive values of principal component four are distinctly associated with the Nueltin granite and there is no clear association of negative PC4 scores with any specific lithology. The linear combinations of elements that represent variability and association within the metric space defined by PCA is governed solely by the stoichiometry of mineralogy, which is only partly dependent on the underlying lithology. In some cases (PC2, PC3, PC4) there is a clear association with principal component scores and specific lithologies. In other cases (PC1, PC5, PC6) there are no obvious associations between lithology and PC scores. Principal component analysis is an effective method for discovering processes that influence the relationships of the variables within geochemical survey data. Dominant eigenvalues followed by a low rate of

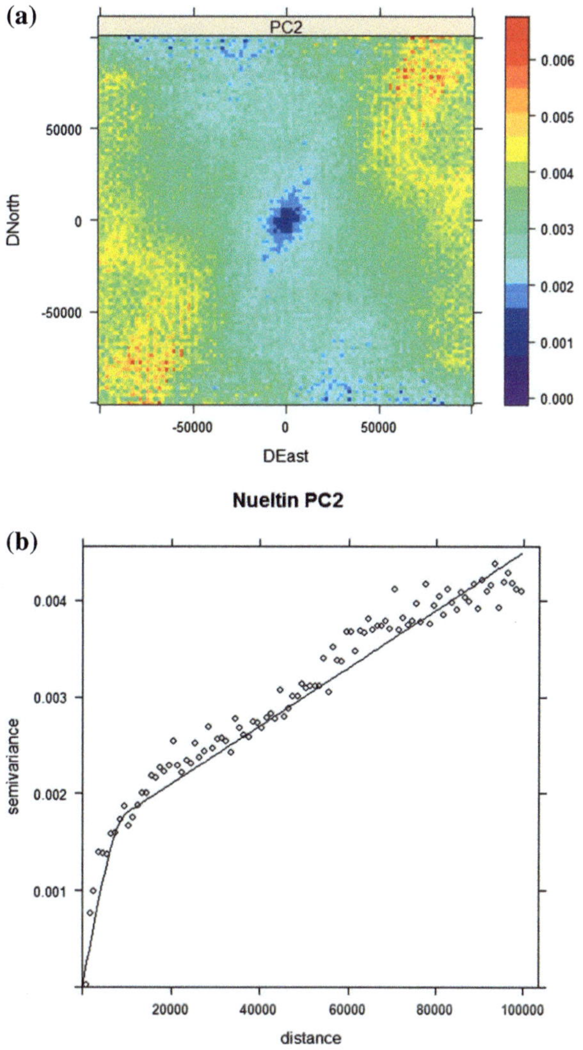

Fig. 8 **a** Variogram map of the second principal component derived from the lake sediment geochemical data. **b** Modeled semivariogram of the second principal component derived from the lake sediment geochemical data. The semivariogram model is derived from a short-range spherical model and a longer range linear model. **c** Interpolated image of the second principal component using the modeled semivariogram shown in Fig. 8b. Grid is in meters. Sites associated with the Nueltin granite show as negative (*dark blue*) areas

decay as exhibited in Fig. 5 demonstrate that there is structure in the data. Given this structure, the application of classification methods can be applied to validate the structure.

3.2.2 Process Validation

The ability to use lake sediment geochemical data for predictive mapping of underlying lithology requires evaluation in the effectiveness of discriminating between the lithologies. To test the predictive map, we apply ANOVA using the logcentered

(c)

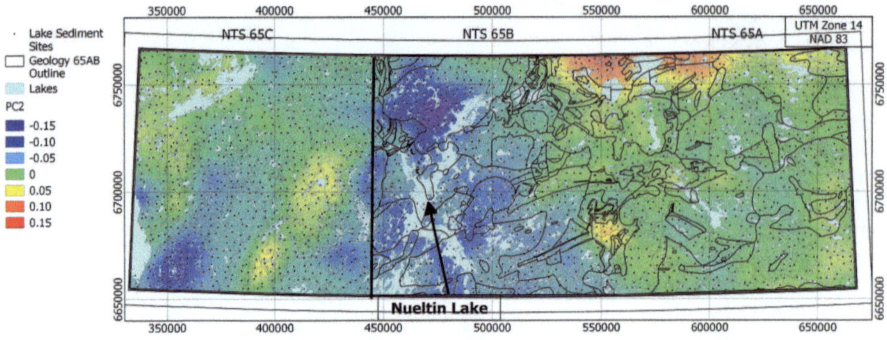

Fig. 8 (continued)

elements, or the principal components derived from the logcentered elements. An ANOVA was carried out on the logcentered data using 10 classes (Ar-c, Ar-dt, Ar-mu, Ar-s, Ar-t, Pp-Hgr, Pp-Hu, Pp-Ng, Pp-Wps, and Pp-Qt) with the elements, and another ANOVA was carried out using the principal components. Figure 9a shows an ordered plot of elements and the associated F-value that provides a measure of class separation. The elements Be–Y–Ni–Ce–La–As–Cr have high F-values and account for much of the lithologic separation (Fig. 9a). ANOVA was also applied to the principal components derived from the logcentered data and Fig. 9b shows a plot of F-value for each principal component. The second principal component accounts for the majority of lithologic separation, followed by PC10, PC14, PC3, PC9, PC18, PC8, PC13, PC7, and PC16 (Fig. 9b). The steep decay of F-values in Fig. 9b suggests that the principal component scores are more effective in the application of classification procedures. The advantage of PC scores derived from logcentered data is that they are orthogonal and represent a linear combination of elements that reflect stoichiometry, thereby making them more realistic in the application of classification methods.

LDA was carried out on ten principal components (PC2, PC10, PC14, PC3, PC9, PC18, PC8, PC13, PC7, and PC16). Cross-validation was used to determine an average accuracy of the classification and the results are shown in Table 3. The first table shows the classification counts for each of the 10 lithologies in a matrix form. The matrix diagonal provides the actual counts that were correctly assigned. The off-diagonal matrix element counts indicate where there is overlap in the classification of the lithologies. These counts are expressed as percentages in the second table. In the second table the classification accuracy for each lithology can be seen along the diagonal and the accuracies range from 28 % (Ar-mu) to 82 % (Ar-c). Both the Hudson granite (Pp-Hgr) and the Nueltin granite (Pp-Ng) show high prediction accuracy (70 and 77 %, respectively). The overall accuracy of the classification is 59 %. The off-diagonal matrix elements of the table indicate the percentage overlap or "confusion" in the classification. It is significant to note that there is

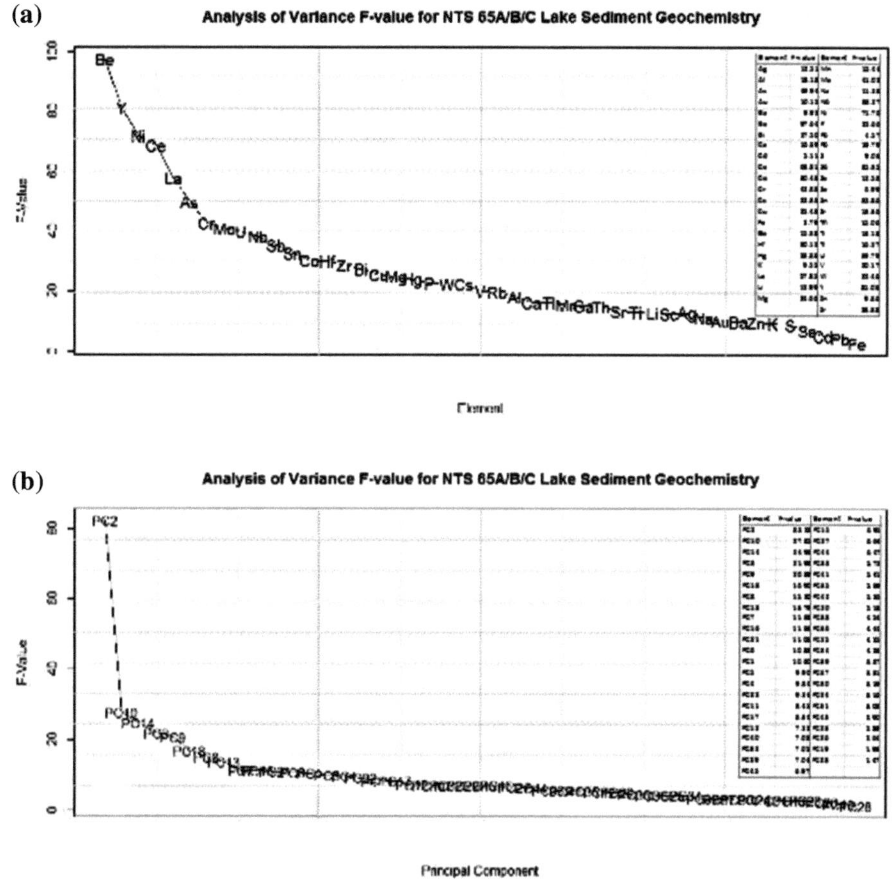

Fig. 9 **a** Ordered plot of F-value and elements as a measure for class separation from the logcentered lake sediment geochemical data. **b** Ordered plot of F-value and principal components as a measure of class separation derived from the logcentered lake sediment geochemical data. The dominance of PC2 over PC1 is reflected in the maps of the principal components. PC2 clearly outlines the two distinct granitic phases (Nueltin and Hudson) whereas these are not recognizable in the map of PC1

little overlap between the Hudson and the Nueltin granites, for example 1.1 % Nu are classified as Hudson, and 0 % Hudson are classified as Nueltin. Both of these granitic rocks show overlap with the Archean tonalite and diorite classes (Ar-t, Ar-dt). These relationships are expressed graphically in Fig. 10 which is a plot of the linear discriminant scores of the sample site observations that are coded according to the initially assigned lithology (symbol) and the lithology assigned from the classification (color). It is clear that many of the sites from NTS65C ("unknown") are classified as Nueltin granite (Pp-Ng), Archean tonalite (Ar-t) and sediment (Ar-s). Convex hull ellipses are drawn for Pp-Ng and Pp-Hgr.

Table 3 Classification accuracy of the lithologies from NTS sheets 65A/B based on linear discriminant analysis of the first seven principal components

	Ar-c	Ar-dt	Ar-mu	Ar-s	Ar-t	Pp-Hgr	Pp-Hu	Pp-Ng	Qt
Ar-c	107	11	0	1	0	11	0	0	1
Ar-dt	23	127	6	9	39	23	3	5	31
Ar-mu	3	10	14	10	5	3	1	0	4
Ar-s	10	5	4	50	14	13	3	0	12
Ar-t	7	34	2	4	133	9	5	14	13
Pp-Hgr	12	24	0	5	8	187	2	0	28
Pp-Hu	0	6	1	5	6	4	22	0	5
Pp-Ng	0	8	0	0	21	2	1	137	9
Qt	1	28	4	10	18	32	5	11	80
	% Accuracy								
	Ar -c	Ar-dt	Ar-mu	Ar-s	Ar-t	Pp-Hgr	Pp-Hu	Pp-Ng	Qt
Ar-c	**81.7**	8.4	0.0	0.8	0.0	8.4	0.0	0.0	0.8
Ar-dt	8.7	**47.7**	2.3	3.4	14.7	8.7	1.1	1.9	11.7
Ar-mu	6.0	20.0	**28.0**	20.0	10.0	6.0	2.0	0.0	8.0
Ar-s	9.0	4.5	3.6	**45.1**	12.6	11.7	2.7	0.0	10.8
Ar-t	3.2	15.4	0.9	1.8	**60.2**	4.1	2.3	6.3	5.9
Pp-Hgr	4.5	9.0	0.0	1.9	3.0	**70.3**	0.8	0.0	10.5
Pp-Hu	0.0	12.2	2.0	10.2	12.2	8.2	**44.9**	0.0	10.2
Pp-Ng	0.0	4.5	0.0	0.0	11.8	1.1	0.6	**77.0**	5.1
Qt	0.5	14.8	2.1	5.3	9.5	16.9	2.7	5.8	**42.3**
Overall Accuracy (%)		58.7							

Posterior probabilities for each of the lithologies were estimated in the application of the linear discriminant analysis. The probability estimates are based on allocation of assignment for each sample site to at least one of the classes based on the Mahalanobis distance to each class multivariate compositional mean based on the 10 principal components used for the analysis.

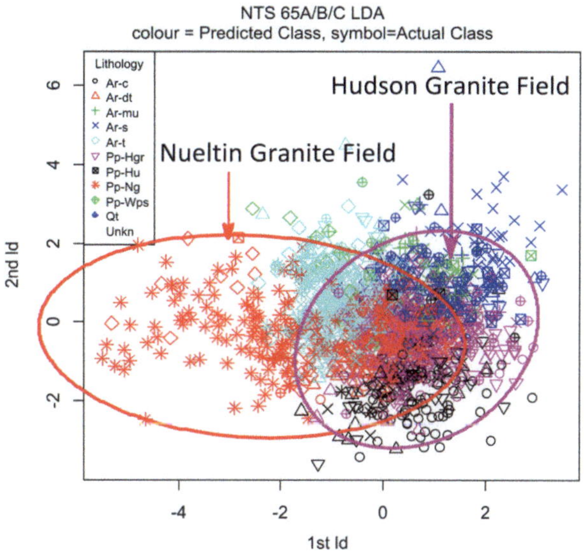

Fig. 10 Plot of linear discriminant function scores 1 and 2 for the lake sediment geochemical data based on selected principal components shown in Fig. 9b. Convex hulls are enclose the Nueltin and Hudson granite values as defined by their initial assignment

Variogram maps and modeled semivariograms (not shown) were created for each of the predicted lithologies. Four of the predicted lithologies, determined by cokriging the posterior probabilities (after an alr transform) and discussed previously, are shown in Fig. 11a–d. Figure 11a displays a predictive map for Archean carbonate rocks in the form of posterior probabilities. The outline of the Archean carbonate lithologies are shown by the black line surrounding posterior probabilities in the southeast portion of the map, but can also be visualized by comparing with the spatial distribution of carbonate rocks as shown in Fig. 4. Posterior probabilities greater than 0.5 are shown on the West side of the map in NTS sheet 65C, where the lithologies are unknown. Figure 11b shows a posterior probability map for Archean tonalite (Ar-t) that has a strong resemblance to the distribution of tonalite in NTS sheets 65 A and 65B, as shown in Fig. 4. An area of increased posterior probability greater than 0.5 also occurs in NTS sheet 65C. Figure 11c show a map of posterior probabilities for the Hudson granite (Pp-Hgr) that closely follows the distribution of the Hudson granite in the East part of the area as shown in Fig. 4. The West portion of the predictive map suggests that the Hudson granite may also be present in NTS 65C. Figure 11d shows a map of posterior probability for the Nueltin granite (Pp-Ng), with high probabilities (greater than 0.8) in NTS 65B, where it closely follows the mapped boundaries (see Fig. 4), as well as a high probability in the southwest part of NTS 65C. Grunsky et al. [14] and Peterson et al. [28] have verified the predicted presence of this lithology in NTS 65C by examining archival bedrock samples.

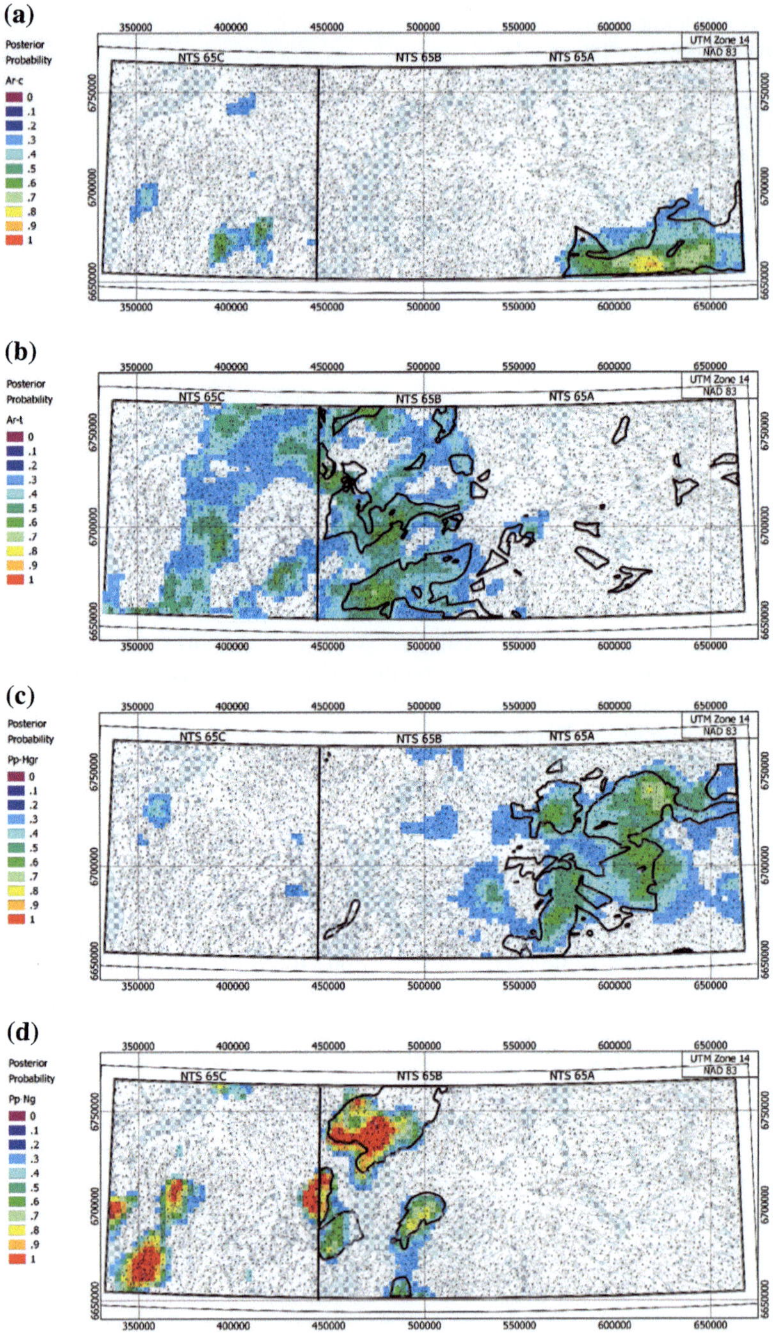
(a)
Posterior Probability
Ar-c
0
.1
.2
.3
.4
.5
.6
.7
.8
.9
1
NTS 65C
NTS 65B
NTS 65A
UTM Zone 14
NAD 83
350000 400000 450000 500000 550000 600000 650000
6650000 6700000 6750000
(b)
Posterior Probability
Ar-t
0
.1
.2
.3
.4
.5
.6
.7
.8
.9
1
NTS 65C
NTS 65B
NTS 65A
UTM Zone 14
NAD 83
350000 400000 450000 500000 550000 600000 650000
6650000 6700000 6750000
(c)
Posterior Probability
Pp-Hgr
0
.1
.2
.3
.4
.5
.6
.7
.8
.9
1
NTS 65C
NTS 65B
NTS 65A
UTM Zone 14
NAD 83
350000 400000 450000 500000 550000 600000 650000
6650000 6700000 6750000
(d)
Posterior Probability
Pp-Ng
0
.1
.2
.3
.4
.5
.6
.7
.8
.9
1
NTS 65C
NTS 65B
NTS 65A
UTM Zone 14
NAD 83
350000 400000 450000 500000 550000 600000 650000
6650000 6700000 6750000

◀ **Fig. 11** **a** Predictive map of Archean carbonate rocks based on cokriging of alr-transformed posterior probabilities derived from the linear discriminant analysis shown in Fig. 10. The partial distribution of Archean carbonate is outlined in *black*. **b** Predictive map of Archean tonalite rocks based on cokriging of alr-transformed posterior probabilities derived from the linear discriminant analysis shown in Fig. 10. The spatial distribution of Archean tonalite is outlined in *black*. **c** Predictive map of Paleoproterozoic Hudson granite based on cokriging of alr-transformed posterior probabilities derived from the linear discriminant analysis shown in Fig. 10. The spatial distribution of Proterozoic Hudson granite is outlined in *black*. **d** Predictive map of Paleoproterozoic Nueltin granite based on cokriging of alr-transformed posterior probabilities derived from the linear discriminant analysis shown in Fig. 10. The spatial distribution of Proterozoic Nueltin granite is outlined in *black*

4 Discussion

The examples presented in this manuscript demonstrate the significance and value of compositional data analysis when applied in conjunction with multivariate statistical procedures and geospatial analysis and presentation. In the case of the Star kimberlite lithogeochemical analyses, the identification of three distinct geochemical trends in the data is illustrated through the combined use of principal component analysis and linear discriminant analysis. Principal components generated from the geochemical data also reflect linear combinations of the elements that are controlled by mineral stoichiometry. This is an important concept in the use of multielement geochemical data for process discovery and validation. The application of this approach for diamond exploration and mining will result in increased efficiencies in both the exploration of, and beneficiation for diamonds.

The lake sediment geochemical survey data from NTS sheets 65A/B/C demonstrate that the use of principal component analysis identifies distinct lithologic differences between the two major Proterozoic granitoid suites (Hudson and Nueltin). The use of principal components derived from the logcentered data for use in classification results in more effective discrimination between the classes. For the lake sediment data, low posterior probabilities can be influenced by significant compositional overlap of the classes, or low initial prior probabilities. Low initial prior probabilities can also indicate a limited geospatial presence in the sampled area. The use of geostatistical procedures to describe, model and display the geospatial features of the predicted lithologies is an important part of the process validation approach. As demonstrated in Figs. 7 and 8, the use of variogram maps and modeled semivariograms can be used to define spatially coherent patterns as shown in the kriged images of the first two principal components. Although not shown, the same methodology was applied to generate the kriged images of Fig. 11a–d. The prediction of the Nueltin granite with a very high posterior probability, in the southwest part of NTS65C, is sufficient evidence that the geochemical uniqueness and spatial coherence of such patterns are valid.

Other variables aside from principal components might also be used including selected isometric log ratio balances or distinct elemental subcompositions. For example, variables derived from the application of multidimensional scaling or independent component analysis (see Venables and Ripley [37]) could be used for process discovery and process validation.

Grunsky [9] provides a suggested methodology for the systematic evaluation of geochemical data. This study uses a subset of that methodology as follows:

Process Discovery

- Examine the marginal distributions of each element with histograms, boxplots, Q–Q plots, scatter plot matrix and summary tables.
- Examine the patterns in geographic space using tools such as a geographic information system.
- Investigate outliers for each element; analytical error, or atypical values? Remove such outliers if necessary.
- Adjust data for censored values if required.
- Apply log-ratio transformations (logcentered, isometric log ratio) so that compositional data can be evaluated without the effect of "closure".
- Tag the geochemical analyses with a categorical variable of interest (lithology, ecosystem, alteration signature, etc.) using a GIS or similar spatial tools to create groups or classes of data.

Process Validation

- Apply ANOVA to the data based on the categorical variables of interest. The ANOVA can be applied to the elements (logratio transformed) or another suitable metric such as principal components.
- Examine the plots of ordered F-value versus variables to determine which variables contribute to maximum discrimination between the groups/classes.
- Apply methods such as linear discriminant analysis to determine the likelihood (probability) of class membership for each observation.
- Generate an accuracy/confusion matrix to indicate the accuracy of prediction for each class and where class overlap occurs.

- Create maps of posterior probability and/or typicality to examine the spatial coherence of the probabilities. In the case of posterior probabilities, the probabilities sum to a constant (1.0) and therefore require a suitable transformation and the use of cokriging with a subsequent back-transformation to create a map of probabilities for each class.

5 Conclusions

The two examples provided in this study provide evidence that the structure of geochemical data is controlled by mineral stoichiometry and when multielement geochemical data are used, the structure of the data is revealed. Multivariate statistical methods, such as principal component analysis that was used in this study shows that distinct lithologies can be identified as part of the discover process in evaluating geochemical data in a compositional paradigm. Classification procedures can take advantage of structure in data to yield classifications that are geochemically distinct and geospatially coherent.

Acknowledgments The authors would like to thank the organizers of the CoDaWork15 meeting for inviting this contribution. In particular the authors wish to thank Santi Thió-Henestrosa and Josep Antoni Martín Fernández of the University of Girona for their support and guidance. Also, discussions and valuable advice from Vera Pawlowsky-Glahn (University of Girona) and Juan Jose Egozcue (Polytechnical University of Catalonia) is gratefully acknowledged. We thank Vera Pawlowsky-Glahn and an anonymous reviewer for their helpful comments, which improved the manuscript.

Appendix A: Summary of the First Seven Principal Components Derived from the Logcentered Lake Sediment Geochemistry

Eigenvalues	PC1	PC2	PC3	PC4	PC5	PC6	PC7
λ	10.63	8.21	4.21	3.17	2.87	2.07	1.58
λ %	23.63	18.25	9.36	7.05	6.38	4.60	3.51
$\Sigma\lambda$ %	23.63	41.89	51.25	58.29	64.67	69.28	72.79

R-Scores	PC1	PC2	PC3	PC4	PC5	PC6	PC7
Ag	-0.6896	-0.3600	-0.0652	-0.4082	-0.0948	-0.0645	0.0484
Al	0.0052	-0.6932	-0.2858	-0.3682	0.2255	-0.0028	0.1103
As	0.0964	0.5020	-0.3098	0.4631	-0.0386	0.1031	0.1433
Au	0.0156	0.2721	0.0718	0.1991	-0.4514	-0.2032	0.5678
Ba	-0.3232	0.3548	-0.2361	-0.0453	0.1789	0.3024	-0.0069
Be	-0.2746	-0.7159	-0.0015	0.1360	0.2290	0.0835	0.0305
Bi	0.2276	-0.3891	0.0465	-0.0911	-0.0228	0.0542	0.3678
Ca	-0.4743	0.6147	0.1971	-0.0488	0.0305	0.3787	-0.1139
Cd	-0.8030	-0.0510	-0.1399	-0.1633	0.0176	0.0774	-0.1275
Ce	-0.3449	-0.6510	0.2431	-0.1650	0.4263	-0.1341	0.1134
Co	0.0241	0.2584	-0.6991	-0.1210	0.1971	-0.2916	-0.1694
Cr	0.4005	0.3444	-0.5225	-0.4324	0.1579	-0.0254	0.1304
Cs	0.6925	-0.4570	0.0460	-0.2453	-0.2714	0.0779	-0.0128
Cu	-0.4470	0.1815	-0.0648	-0.5171	-0.2777	-0.4290	0.0366
Fe	0.0820	-0.1493	-0.5839	0.3819	0.4700	-0.1869	-0.1537
Ga	0.7319	-0.2796	-0.1433	-0.4141	0.1087	0.1728	0.0884
Hf	0.3365	0.6999	0.3523	0.0863	0.2250	-0.1225	-0.1581
Hg	-0.5653	0.4059	0.0649	-0.2984	0.0992	0.0415	-0.0768
K	0.8470	-0.1086	-0.0075	-0.2016	-0.2658	0.1034	-0.1063
La	-0.3857	-0.6293	0.3271	-0.1658	0.3479	-0.1474	0.1056
Li	0.7502	-0.4464	0.0044	-0.2162	-0.2853	0.0690	-0.0848
Mg	0.6934	0.4307	-0.1744	-0.1494	-0.1714	0.2361	-0.0460
Mn	-0.0738	-0.0796	-0.5872	0.5283	0.0159	0.1597	-0.0231
Mo	-0.4525	-0.5731	0.0191	0.2718	-0.1175	-0.0892	-0.3368
Na	0.0236	0.0595	0.1551	-0.2829	-0.2194	0.4713	-0.4132
Nb	0.4089	-0.0611	0.4859	-0.3059	0.3879	0.0059	-0.1595
Ni	-0.0469	0.4945	-0.4123	-0.4900	-0.0553	-0.3776	0.0277
P	-0.2586	-0.4128	-0.5709	0.0324	0.0914	0.4426	0.1062
Pb	0.0266	-0.0242	0.2412	-0.1052	0.5382	0.2861	0.2982
Rb	0.8060	-0.3750	0.1331	-0.2019	-0.2013	0.0807	-0.0855
S	-0.6413	0.3283	0.2189	-0.2093	-0.1775	-0.0567	-0.3736
Sb	-0.3693	0.5554	0.0347	-0.0039	-0.1051	0.2420	0.0315
Sc	0.6690	0.1889	-0.0053	0.1187	0.3886	-0.2306	-0.2066
Se	-0.7685	0.0353	0.0363	-0.1865	-0.0536	0.0686	-0.1227
Sn	0.6104	-0.4045	0.2635	-0.0518	0.1928	0.3464	-0.0125
Sr	-0.3994	0.5417	0.1022	-0.2750	0.4482	0.2338	0.0405
Th	0.5716	-0.0714	0.3269	0.4279	0.3280	-0.3448	-0.1567

Ti	0.9094	0.1670	0.0073	-0.1135	0.1189	0.0252	-0.1138
Tl	0.3195	-0.5839	-0.1892	-0.0986	-0.2403	-0.2824	-0.2146
U	-0.2758	-0.4211	0.4373	0.2342	-0.3026	-0.0621	-0.0397
V	0.3048	-0.0219	-0.6501	-0.1289	0.3147	0.0346	-0.0835
W	-0.1031	-0.4562	-0.1866	0.3853	-0.1297	0.2744	-0.2798
Y	-0.4994	-0.6529	0.2535	-0.0544	0.2755	-0.0378	0.0049
Zn	-0.4385	-0.4681	-0.3566	-0.1699	-0.0125	-0.0915	-0.2943
Zr	0.3283	0.7017	0.4047	0.0199	0.2670	-0.1594	-0.1600

Relative Contributions

	PC1	PC2	PC3	PC4	PC5	PC6	PC7
Ag	47.5809	12.9652	0.4254	16.6696	0.8983	0.4168	0.2346
Al	0.0027	48.0709	8.1692	13.5636	5.0854	0.0008	1.2181
As	0.9304	25.2093	9.6036	21.4559	0.1488	1.0626	2.0554
Au	0.0243	7.4055	0.5161	3.9643	20.3829	4.1312	32.2588
Ba	10.4501	12.5913	5.5757	0.2049	3.2023	9.1515	0.0047
Be	7.5460	51.2669	0.0002	1.8513	5.2475	0.6982	0.0930
Bi	5.1816	15.1480	0.2164	0.8298	0.0520	0.2944	13.5354
Ca	22.5056	37.8040	3.8882	0.2382	0.0930	14.3466	1.2989
Cd	64.5047	0.2603	1.9584	2.6671	0.0309	0.5992	1.6258
Ce	11.8979	42.4009	5.9103	2.7225	18.1834	1.7998	1.2867
Co	0.0582	6.6789	48.8894	1.4643	3.8847	8.5044	2.8721
Cr	16.0471	11.8655	27.3116	18.7029	2.4933	0.0643	1.7021
Cs	47.9793	20.8972	0.2121	6.0197	7.3714	0.6069	0.0165
Cu	19.9867	3.2940	0.4201	26.7503	7.7156	18.4129	0.1338
Fe	0.6720	2.2290	34.1040	14.5916	22.0995	3.4942	2.3621
Ga	53.5866	7.8207	2.0547	17.1519	1.1811	2.9856	0.7809
Hf	11.3276	49.0005	12.4140	0.7444	5.0637	1.5008	2.5013
Hg	31.9687	16.4852	0.4211	8.9087	0.9847	0.1727	0.5906
K	71.7801	1.1798	0.0056	4.0661	7.0695	1.0704	1.1295
La	14.8804	39.6189	10.7020	2.7486	12.1102	2.1722	1.1165
Li	56.3047	19.9365	0.0019	4.6778	8.1438	0.4759	0.7200
Mg	48.1002	18.5566	3.0432	2.2334	2.9397	5.5783	0.2119
Mn	0.5443	0.6341	34.5010	27.9216	0.0254	2.5506	0.0534
Mo	20.4828	32.8594	0.0366	7.3922	1.3819	0.7954	11.3483
Na	0.0558	0.3543	2.4078	8.0043	4.8150	22.2186	17.0774
Nb	16.7311	0.3733	23.6240	9.3641	15.0570	0.0035	2.5462
Ni	0.2201	24.4641	17.0040	24.0195	0.3059	14.2657	0.0767
P	6.6900	17.0471	32.6053	0.1050	0.8358	19.5941	1.1274
Pb	0.0710	0.0588	5.8193	1.1079	28.9736	8.1893	8.8933

Rb	64.9861	14.0714	1.7719	4.0761	4.0526	0.6510	0.7311
S	41.1495	10.7839	4.7924	4.3827	3.1523	0.3215	13.9660
Sb	13.6422	30.8653	0.1203	0.0016	1.1053	5.8572	0.0996
Sc	44.7749	3.5709	0.0028	1.4102	15.1108	5.3222	4.2716
Se	59.0890	0.1249	0.1319	3.4799	0.2870	0.4708	1.5067
Sn	37.2800	16.3731	6.9460	0.2681	3.7187	12.0059	0.0157
Sr	15.9626	29.3557	1.0451	7.5633	20.0953	5.4695	0.1639
Th	32.6887	0.5099	10.6924	18.3150	10.7648	11.8954	2.4556
Ti	82.7357	2.7906	0.0053	1.2893	1.4152	0.0637	1.2961
Tl	10.2129	34.1041	3.5828	0.9716	5.7767	7.9797	4.6067
U	7.6113	17.7408	19.1341	5.4896	9.1582	0.3852	0.1575
V	9.2929	0.0481	42.2819	1.6611	9.9078	0.1199	0.6981
W	1.0627	20.8222	3.4840	14.8545	1.6826	7.5351	7.8332
Y	24.9539	42.6497	6.4278	0.2956	7.5932	0.1426	0.0024
Zn	19.2395	21.9176	12.7235	2.8888	0.0156	0.8381	8.6661
Zr	10.7850	49.2607	16.3838	0.0398	7.1294	2.5410	2.5620

Absolute Contributions

	PC1	PC2	PC3	PC4	PC5	PC6	PC7
Ag	4.4737	1.5783	0.1010	5.2564	0.3133	0.2016	0.1486
Al	0.0003	5.8518	1.9387	4.2770	1.7735	0.0004	0.7714
As	0.0875	3.0688	2.2791	6.7657	0.0519	0.5139	1.3017
Au	0.0023	0.9015	0.1225	1.2500	7.1085	1.9981	20.4294
Ba	0.9825	1.5328	1.3232	0.0646	1.1168	4.4262	0.0030
Be	0.7095	6.2409	0.0001	0.5838	1.8300	0.3377	0.0589
Bi	0.4872	1.8440	0.0514	0.2617	0.0181	0.1424	8.5719
Ca	2.1160	4.6020	0.9228	0.0751	0.0324	6.9389	0.8226
Cd	6.0649	0.0317	0.4648	0.8410	0.0108	0.2898	1.0296
Ce	1.1187	5.1616	1.4027	0.8585	6.3414	0.8705	0.8149
Co	0.0055	0.8130	11.6026	0.4617	1.3548	4.1133	1.8189
Cr	1.5088	1.4444	6.4817	5.8976	0.8695	0.0311	1.0779
Cs	4.5111	2.5439	0.0503	1.8982	2.5707	0.2935	0.0105
Cu	1.8792	0.4010	0.0997	8.4352	2.6908	8.9056	0.0847
Fe	0.0632	0.2713	8.0937	4.6012	7.7071	1.6900	1.4959
Ga	5.0383	0.9520	0.4876	5.4085	0.4119	1.4440	0.4946
Hf	1.0650	5.9650	2.9461	0.2347	1.7660	0.7259	1.5841
Hg	3.0058	2.0068	0.0999	2.8092	0.3434	0.0835	0.3740
K	6.7489	0.1436	0.0013	1.2822	2.4654	0.5177	0.7153
La	1.3991	4.8230	2.5398	0.8667	4.2234	1.0506	0.7071
Li	5.2939	2.4269	0.0005	1.4750	2.8401	0.2302	0.4560

Mg	4.5225	2.2590	0.7222	0.7042	1.0252	2.6980	0.1342
Mn	0.0512	0.0772	8.1879	8.8045	0.0089	1.2336	0.0338
Mo	1.9258	4.0001	0.0087	2.3310	0.4819	0.3847	7.1868
Na	0.0053	0.0431	0.5714	2.5240	1.6792	10.7463	10.8151
Nb	1.5731	0.0454	5.6065	2.9528	5.2511	0.0017	1.6125
Ni	0.0207	2.9781	4.0354	7.5740	0.1067	6.8998	0.0486
P	0.6290	2.0752	7.7380	0.0331	0.2915	9.4769	0.7140
Pb	0.0067	0.0072	1.3811	0.3493	10.1044	3.9609	5.6321
Rb	6.1101	1.7130	0.4205	1.2853	1.4133	0.3149	0.4630
S	3.8690	1.3128	1.1374	1.3820	1.0994	0.1555	8.8446
Sb	1.2827	3.7573	0.0285	0.0005	0.3855	2.8329	0.0631
Sc	4.2098	0.4347	0.0007	0.4447	5.2698	2.5741	2.7052
Se	5.5557	0.0152	0.0313	1.0973	0.1001	0.2277	0.9542
Sn	3.5052	1.9932	1.6484	0.0845	1.2969	5.8068	0.0100
Sr	1.5008	3.5736	0.2480	2.3849	7.0082	2.6454	0.1038
Th	3.0735	0.0621	2.5375	5.7753	3.7542	5.7533	1.5551
Ti	7.7790	0.3397	0.0012	0.4066	0.4935	0.0308	0.8208
Tl	0.9602	4.1516	0.8503	0.3064	2.0146	3.8595	2.9174
U	0.7156	2.1597	4.5410	1.7310	3.1939	0.1863	0.0998
V	0.8737	0.0059	10.0345	0.5238	3.4553	0.0580	0.4421
W	0.0999	2.5348	0.8268	4.6841	0.5868	3.6444	4.9607
Y	2.3462	5.1919	1.5255	0.0932	2.6481	0.0690	0.0015
Zn	1.8089	2.6681	3.0196	0.9109	0.0054	0.4053	5.4882
Zr	1.0140	5.9967	3.8882	0.0125	2.4864	1.2290	1.6225

References

1. Aitchison, J.: The Statistical Analysis of Compositional Data (Reprinted in 2003 by The Blackburn Press), p. 416. Chapman & Hall Ltd., London (1986)
2. Aitchison, J.: Logratios and natural laws in compositional data analysis. Math. Geol. **31**(5), 563–580 (1999)
3. Bivand, R., Pebesma, E., Gomez-Rubio, V.: Applied Spatial Data Analysis with R, 2nd edn, 405pp. Springer (2013)
4. Buccianti, A., Mateu-Figueras, G., Pawlowsky-Glahn, V.: Compositional data analysis in the geosciences: from theory to practice. Geol. Soc. Spec. Publ. **264**, 212p (2006)
5. Eade, K.E.: Geology, Nueltin Lake, District of Keewatin, Geological Survey of Canada. Preliminary Map 4-1972, 1 sheet (1973a). doi:10.4095/108984
6. Eade, K.E.: Edehon Lake Area, West Half, District of Keewatin, Geological Survey of Canada. Preliminary Map 3-1972, 1973, 1 sheet (1973b). doi:10.4095/108978
7. Egozcue, J.J., Pawlowsky-Glahn, V., Mateu-Figueras, G., Barcelo-Vidal, C.: Isometric Logratio transformations for compositional data analysis. Math. Geol. **35**, 279–300 (2003)
8. Grunsky, E.C.: A program for computing RQ-mode principal components analysis for S-plus and R. Comput. Geosci. **27**, 229–235 (2001)

9. Grunsky, E.C.: The interpretation of geochemical survey data. Geochem. Explor. Environ. Anal. **10**, 27–74 (2010)
10. Grunsky, E.C.: Predicting archean volcanogenic massive sulfide deposit potential from lithogeochemistry: application to the Abitibi greenstone belt. Geochem. Explor. Environ. Anal. **13**(2013), 317–336 (2013). doi:10.1144/geochem2012-176
11. Grunsky, E.C., Kjarsgaard, B.A.: Classification of eruptive phases of the Star Kimberlite, Saskatchewan, Canada based on statistical treatment of whole-rock geochemical analyses. Appl. Geochem. **23**(12), 3321–3336 (2008). (ESS Contribution # 20080330)
12. Grunsky, E.C., Bacon-Shone, J.: The stoichiometry of mineral compositions. In: Proceedings of the 4th International Workshop on Compositional Data Analysis. Sant Feliu de Guixols, Spain (2011)
13. Grunsky, E.C., Corrigan, D., Mueller, U., Bonham-Carter. G-F.: Predictive geologic mapping using lake sediment geochemistry in the Melville Peninsula Geological Survey of Canada, Open File 7171, 1 sheet (2012a). doi:10.4095/291901
14. Grunsky, E.C., McCurdy, M.W., Pehrsson, S.J., Peterson, T.D., Bonham-Carter, G.F.: Predictive geologic mapping and assessing the mineral potential in NTS 65A/B/C, Nunavut, with new regional lake sediment geochemical data; Geological Survey of Canada, Open File 7175, 1 sheet (2012b). doi:10.4095/291920
15. Grunsky, E.C., Mueller, U.A., Corrigan, D.: A study of the lake sediment geochemistry of the Melville Peninsula using multivariate methods: applications for predictive geological mapping. J. Geochem. Explor. **141**, 15–41 (2014). doi:10.1016/j.gexplo.2013.07.013
16. Hron, K., Templ, M., Filzmoser, P.: Imputation of missing values for compositional data using classical and robust methods. Comput. Stat. Data Anal. **54**(12), 3095–3107 (2010)
17. Martín-Fernández, J.A., Barceló-Vidal, C., Pawlowsky-Glahn, V.: Dealing with zeros and missing values in compositional data sets using nonparametric imputation. Math. Geol. **35**(3), 253–278 (2003)
18. Martín-Fernández, J.A., Palarea, J., Olea, R.: Dealing with Zeros, pp. 43-58j, 378 p. Wiley (2011)
19. McCurdy, M.W., McNeil, R.J., Day, S.J.A., Pehrsson, S.J.: Regional lake sediment and water geochemical data, Nueltin Lake area, Nunavut (NTS 65A, 65B and 65C), Geological Survey of Canada, Open File 6986, 13 pp (2012) 1 CD-ROM. doi:10.4095/289888
20. Palarea-Albaladejo, J., Martín-Fernández, J.A.: A modified EM alr-algorithm for replacing rounded zeros in compositional data sets. Comput. Geosci. **34**(8), 902–917 (2008)
21. Palarea-Albaladejo, J., Martín-Fernández, J.A., Buccianti, A.: Compositional methods for estimating elemental concentrations below the limit of detection in practice using R. J. Geochem. Explor. **141**, 71–77 (2014). doi:10.1016/j.gexplo.2013.09.003
22. Pawlowsky-Glahn, V., Buccianti, A. (eds.): Compositional Data Analysis: Theory and Application. Wiley, New York (2011)
23. Pawlowsky-Glahn, V., Egozcue, J.J.: Spatial analysis of compositional data: a historical review. J. Geochem. Explor. (2016). doi:10.1016/j.gexplo.2015.12.010
24. Pawlowsky-Glahn, V., Olea, R.A.: Geostatistical Analysis of Compositional Data, Studies in Mathematical Geology, vol. 7, 181 p. Oxford University Press
25. Pearce, T.H.: A contribution to the theory of variation diagrams. Contrib. Mineral. Petrol. **19**(2), 142–157 (1968)
26. Pebesma, E.J.: Multivariable geostatistics in S: the gstat package. Comput. Geosci. **30**, 683–691 (2004)
27. Peterson, T.D., Scott, J.M.J., Jefferson, C.W., Tschirhart, V.: Regional potassic alteration corridors spatially related to the 1750 Ma Nueltin Suite in the northeast Thelon Basin region, Nunavut—guides to uranium, gold and silver? In: Geological Association of Canada-Mineralogical Association of Canada, Joint Annual Meeting, Programs with Abstracts, vol. 35, p. 1 (2012)
28. Peterson, T.D., Scott, J.M.J., Lecheminant, A.N., Chorlton, L.B., D'Aoust, B.M.A.: Geological Survey of Canada, Canadian Geoscience Map 158, 1 sheet (2014). doi:10.4095/293892

29. Peterson, T.D., Scott, J.M.J., LeCheminant, A.N., Jefferson, C.W., Pehrsson, S.J.: The Kivalliq igneous suite: anorogenic bimodal magmatism at 1.75 Ga in the western Churchill Province, Canada. Precambr. Res. **262**, 101–119 (2015). http://dx.doi.org/10.1016/j.precamres.2015.02.019
30. QGIS Development Team: QGIS Geographic Information System. Version 2.8.1-Wien. Open Source Geospatial Foundation (2015). http://qgis.osgeo.org
31. R Core Team: R: A language and environment for statistical computing. R Foundation for Statistical Computing, Vienna, Austria (2014). http://www.R-project.org/
32. Stanley, C.R.: Effects of non-conserved denominators on Pearce element ratio diagrams. Math. Geol. **25**(8), 1049–1070 (1993)
33. Tella, S., Paul, D., Berman, R.G., Davis, W.J., Peterson, T.D., Pehrsson, S.J., Kerswill, J.A.: Geological Survey of Canada, Open File 5441, 3 sheets (2007) 1 CD-ROM. doi:10.4095/224573
34. Tolosana-Delgado, R.: Geostatistics for constrained variables: positive data, compositions and probabilities. Applications to environmental hazard monitoring, Ph.D. Thesis, University of Girona, 215p (2006)
35. Urqueta, E., Kyser, T.K., Clark, A.H., Stanley, C.R., Oates, C.J.: Lithogeochemistry of the collahuasi porphyry Cu-Mo and epithermal Cu-Ag (-Au) cluster, northern Chile: pearce element ratio vectors to ore. Geochem. Explor. Environ. Anal. **9**(1), 9–17 (2009)
36. van Breemen, O., Peterson, T.D., Sandeman, H.A.: U-Pb zircon geochronology and Nd isotope geochemistry of proterozoic granitoids in the western Churchill Province: intrusive age pattern and Archean source domains. Can. J. Earth Sci. **42**, 339–377 (2005)
37. Venables, W.N., Ripley, B.D.: Modern Applied Statistics with S, 4th edn, 495 p. Springer, Berlin (2002)

Space-Time Compositional Models: An Introduction to Simplicial Partial Differential Operators

E. Jarauta-Bragulat and J.J. Egozcue

Abstract A function assigning a composition to space-time points is called a compositional or simplicial field. These fields can be analyzed using the compositional analysis tools. In order to study compositions depending on space and/or time, reformulation and interpretation of traditional partial differential operators is required. These operators such as: partial derivatives, compositional gradient, directional derivative and divergence are of primary importance to state alternative models of processes as diffusion, advection and waves, from the compositional perspective. This kind of models, usually based on continuity of mass, circulation of a vector field along a curve and flux through surfaces, should be analyzed when compositional operators are used instead of the traditional gradient or divergence. This study is aimed at setting up the definitions, mathematical basis and interpretation of such operators.

Keywords Compositional derivative · Aitchison geometry · Mass continuity · Gradient · Gauss divergence theorem

1 Introduction

In a large number of processes studied in Sciences and in Engineering, magnitudes or variables involved can be modelled by a vector. This vector may be a function of one or several variables. Furthermore, in many cases the studied vector is a composition.

A study of linear models for evolutionary compositions depending on one variable, usually time, was formulated by Egozcue and Jarauta-Bragulat [3] in terms of the so-called simplicial linear differential equations. The foundations of differential and integral calculus for simplex-valued functions of one real variable, was presented

E. Jarauta-Bragulat (✉) · J.J. Egozcue
Department of Enginyeria Civil I Ambiental, Universitat Politècnica de Catalunya,
C/ Jordi Girona, 1-3 (C2-ETSECCPB-UPC), 08034 Barcelona, Spain
e-mail: eusebi.jarauta@upc.edu

J.J. Egozcue
e-mail: juan.jose.egozcue@upc.edu

J.A. Martín-Fernández and S. Thió-Henestrosa (eds.), *Compositional Data Analysis*, Springer Proceedings in Mathematics & Statistics 187,
DOI 10.1007/978-3-319-44811-4_8

by Egozcue et al. [4]. The present work focuses on compositions whose evolution depends on several variables; in many cases, these variables are spatial coordinates and time. For example, the study of the evolution of a pollutant carried by a fluid stream.

For spatial coordinates, a subset $S \in \mathbb{R}^d$ of a d-dimensional real space is considered and identified with a physical domain; consequently $d = 1$, $d = 2$ and $d = 3$ are the common choices of that dimension. A location in this space is denoted $s \in S$ and represented in a Cartesian coordinate system; for $d = 1$ the spatial coordinate is usually denoted $s = x$; for $d = 2$, $s = (x, y)$ and so on. For time, a subset $T \subseteq \mathbb{R}$ is considered and a point is denoted $t \in T$. For spatial and time evolutionary processes, a domain $S \times T \subseteq \mathbb{R}^d \times \mathbb{R}$ $(d = 1, 2, 3)$ is considered.

A space-time vector-valued field with positive components $\mathbf{Z}$ is a function $\mathbf{Z} : S \times T \subseteq \mathbb{R}^d \times \mathbb{R} \to \mathbb{R}^n_+$ that assigns a positive component vector $\mathbf{Z}(s, t) \in \mathbb{R}^n_+$ to a space-time point $(s, t) \in S \times T$. If closure operation is then applied: $\mathcal{C}\mathbf{Z}(s, t) = \mathbf{z}(s, t)$ a space-time simplicial field $\mathbf{z}$ (STSF) is obtained. Consequently, a STSF is a function $\mathbf{z} : S \times T \subseteq \mathbb{R}^d \times \mathbb{R} \to \mathbb{S}^n$, that assigns a composition $\mathbf{z}(s, t)$ in the n-part simplex $\mathbb{S}^n$ to any space-time point $(s, t) \in S \times T$. Throughout this paper, the existence of derivatives or integrals of any field is assumed.

2 Derivatives and Integrals of a Space-Time Simplicial Field

In the following, definitions and properties are developed for $d = 2$ and extension to other values of d are natural. Based on the definition of (ordinary) simplicial derivative [4] and the real calculus, natural definitions for partial simplicial derivatives of a STSF follow.

Definition 1 (*Spatial and time simplicial derivatives*) Let $\mathbf{z} : S \times T \subseteq \mathbb{R}^2 \times \mathbb{R} \to \mathbb{S}^n$ be a STSF and $(x, y) \in S$. The spatial-partial simplicial derivatives of $\mathbf{z}$ are

$$\partial_x^{\oplus}\mathbf{z}(x, y, t) = \lim_{h \to 0} \left(\frac{1}{h} \odot (\mathbf{z}(x + h, y, t) \ominus \mathbf{z}(x, y, t)) \right) ,$$

$$\partial_y^{\oplus}\mathbf{z}(x, y, t) = \lim_{h \to 0} \left(\frac{1}{h} \odot (\mathbf{z}(x, y + h, t) \ominus \mathbf{z}(x, y, t)) \right) .$$

The time-partial simplicial derivative is

$$\partial_t^{\oplus}\mathbf{z}(x, y, t) = \lim_{h \to 0} \left(\frac{1}{h} \odot (\mathbf{z}(x, y, t + h) \ominus \mathbf{z}(x, y, t)) \right) .$$

Definition 2 (*Directional simplicial derivatives*) Let $\mathbf{z} : S \times T \subseteq \mathbb{R}^2 \times \mathbb{R} \to \mathbb{S}^n$ be a STSF. Let $(x, y) \in S$ and $\mathbf{u} = (u_x, u_y)$ be a vector in $\mathbb{R}^2$. The simplicial derivative of $\mathbf{z}$ with respect to $\mathbf{u}$ is

$$\partial_{\mathbf{u}}^{\oplus}\mathbf{z}(x,y,t) = \lim_{h\to 0}\left(\frac{1}{h}\odot\left(\mathbf{z}(x+hu_x, y+hu_y, t)\ominus\mathbf{z}(x,y,t)\right)\right).$$

If $\mathbf{u}$ is a unit vector, the derivative is called the directional simplicial derivative of $\mathbf{z}$.

Spatial-partial and time-partial simplicial derivatives are computed as they were ordinary simplicial derivatives of a single variable simplex-valued function as developed in Egozcue et al. [4]. Consequently, they can be computed as ordinary derivatives of the log transformation of the STSF, and then transformed back into compositions. The same scheme works for clr and ilr transformations of $\mathbf{z}$ [1, 5]. The following proposition summarize this kind of computation.

Proposition 1 *Partial simplicial derivatives of* $\mathbf{z}(x,y,t)$ *can be computed as*

$$\partial_x^{\oplus}\mathbf{z}(x,y,t) = \mathcal{C}\exp\left(\partial_x\log(\mathbf{z}(x,y,t))\right) = \mathcal{C}\exp\begin{pmatrix}\frac{\partial_x z_1(x,y,t)}{z_1(x,y,t)}\\ \vdots \\ \frac{\partial_x z_n(x,y,t)}{z_n(x,y,t)}\end{pmatrix}.$$

Additionally

$$\mathrm{clr}\left(\partial_x^{\oplus}\mathbf{z}(x,y,t)\right) = \partial_x\mathrm{clr}(\mathbf{z}(x,y,t));\quad \mathrm{ilr}\left(\partial_x^{\oplus}\mathbf{z}(x,y,t)\right) = \partial_x\mathrm{ilr}(\mathbf{z}(x,y,t)).$$

Similar expressions hold for $\partial_y^{\oplus}\mathbf{z}(x,y,t)$ *and* $\partial_t^{\oplus}\mathbf{z}(x,y,t)$.

Let $\mathbf{f}: I\subseteq\mathbb{R}\to\mathbb{S}^n$ be a continuous simplex-valued function of real variable; a differentiable function $\mathbf{F}: I\subseteq\mathbb{R}\to\mathbb{S}^n$ is a *simplicial antiderivative* of $\mathbf{f}$ on I if, and only if, $\partial^{\oplus}\mathbf{F}(\xi) = \mathbf{f}(\xi)$, $\xi\in I$. Other definitions and properties related to integral of simplex-valued functions of one variable have been stated in Egozcue et al. [4]. Some of them are summarised in the following proposition:

Proposition 2 *Let* $\mathbf{f}: I\subseteq\mathbb{R}\to\mathbb{S}^n$ *be a continuous simplex-valued function of real variable.*

- $\int^{\oplus} d\xi\odot\mathbf{f}(\xi) = \mathcal{C}\exp\left(\int\log(\mathbf{f}(\xi))d\xi\right)$.
- $\int^{\oplus}_{[a,b]} d\xi\odot\mathbf{f}(\xi) = \mathcal{C}\exp\left(\int_a^b\log(\mathbf{f}(\xi))d\xi\right)$.
- $\mathrm{clr}\left(\int^{\oplus}_{[a,b]} d\xi\odot\mathbf{f}(\xi)\right) = \int_a^b\mathrm{clr}(\mathbf{f}(\xi))d\xi$.
- $\mathrm{ilr}\left(\int^{\oplus}_{[a,b]} d\xi\odot\mathbf{f}(\xi)\right) = \int_a^b\mathrm{ilr}(\mathbf{f}(\xi))d\xi$.

Some natural extended definitions and properties can be stated for double and line integrals.

Definition 3 (*Double integrals*) Let $\mathbf{z}: S\times T\subseteq\mathbb{R}^2\times\mathbb{R}\to\mathbb{S}^n$ be a STSF. The double integral of $\mathbf{z}$ in S is a composition in $\mathbb{S}^n$ given by

$$\iint_S^{\oplus}(d\xi d\eta)\odot\mathbf{z}(\xi,\eta,t) = \mathcal{C}\exp\left(\iint_S\log(\mathbf{z}(\xi,\eta,t))\,d\xi d\eta\right).$$

Proposition 3 *Properties of double integrals are*

$$\mathrm{clr}\left(\iint_S^{\oplus} (d\xi d\eta) \odot \mathbf{z}(\xi,\eta,t)\right) = \iint_S \mathrm{clr}(\mathbf{z}(\xi,\eta,t)) d\xi d\eta \;,$$

$$\mathrm{ilr}\left(\iint_S^{\oplus} (d\xi d\eta) \odot \mathbf{z}(\xi,\eta,t)\right) = \iint_S \mathrm{ilr}(\mathbf{z}(\xi,\eta,t)) d\xi d\eta \;.$$

Definition 4 (*Line integrals*) Let $\mathbf{z} : S \times T \subseteq \mathbb{R}^2 \times \mathbb{R} \to \mathbb{S}^n$ be a STSF. The line integral of $\mathbf{z}$ along a regular curve of finite length Γ associated with the function $s : [a,b] \subset \mathbb{R} \to S$, is

$$\int_{\Gamma}^{\oplus} ds \odot \mathbf{z}(x(u), y(u), t) = \mathcal{C}\exp\left(\int_a^b \log(\mathbf{z}(x(u), y(u), t))\, s'(u)\, du\right) \;,$$

where $\gamma_1 = (x(a), y(a))$, $\gamma_2 = (x(b), y(b))$ are the end points of the curve Γ and $s'(u)$ denotes the ordinary derivative of s with respect to the parameter u.

Proposition 4 *Properties of line integrals are:*

$$\mathrm{clr}\left(\int_{\Gamma}^{\oplus} (ds \odot \mathbf{z}(x(u), y(u), t))\right) = \int_a^b \mathrm{clr}(\mathbf{z}(x(u), y(u), t))\, s'(u)\, du \;,$$

$$\mathrm{ilr}\left(\int_{\Gamma}^{\oplus} (ds \odot \mathbf{z}(x(u), y(u), t))\right) = \int_a^b \mathrm{ilr}(\mathbf{z}(x(u), y(u), t))\, s'(u)\, du \;.$$

3 Compositional Mass Continuity Equation and Differential Operators

When n species are mixed in a continuum, a common assumption is that the mass of each species changes according the input–output of it through the border of a fixed control volume V. When working in a space of dimension $d = 2$, volume means area, or alternatively, for $d = 1$ is just a length. Notation V is used both for referring to the volume itself and for indicating its magnitude in some volume unit. The continuity of mass for each species is normally described by the continuity equation [6, 8]. It can be written as

$$\partial_t \rho_k + \mathrm{div}(\rho_k \mathbf{v}_k) = 0, \quad k = 1, 2, \ldots n \;, \tag{1}$$

where ρ_k is the mass density of the k-species, and $\mathbf{v}_k = (v_{kx}, v_{ky})$ is its velocity in a planar movement. Attention should be paid to the definition of ρ_k. It is the ratio of the mass m_k to the volume V_k occupied by the mass of the k-species. Accordingly, $\rho_k = m_k/V_k$ and the units can be, for instance, g/cm^3. Interest is centred in the behaviour of (mass) concentration $c_k = m_k/M$ of each species, which is given as the ratio of the mass of the k-species to the total mass $M = \sum m_k$ within some given control volume V. Note that depending on the assumptions about the material, V_k may be equal to V (perfect gasses) or not. For instance, $V = \sum V_k$ for solid materials or non reactive liquids. The overall density is $\rho = M/V$ which leads to a convenient expression of ρ_k in terms of concentrations

$$\rho_k = \frac{m_k}{V_k} = \frac{m_k/M}{V_k/\rho V} = \rho \frac{m_k/M}{V_k/V} = \rho \frac{c_k}{a_k} = \rho d_k, \quad k = 1, 2, \dots, n \ , \tag{2}$$

where $a_k = V_k/V$ is the volume fraction or volume concentration of the k-species. If $V_k = V$ for all the species, then $a_k = 1$ in all cases. The ratio of mass concentration to volume concentration is denoted $d_k = c_k/a_k$. Note that the continuity equations hold for each species but not for the overall density ρ, as the change of concentrations modifies the mass content of V and the diffusion or selective transport of some species may change the overall density ρ. Continuity equation (1) does not hold for the overall density ρ if ρ and $\mathbf{v}_k$ are different and space-time dependent. For a planar flow, the fields of densities are considered functions of space location $s \in S \subseteq \mathbb{R}^2$ and time $t \in T \subseteq \mathbb{R}$. The explicit dependence is suppressed unless it is necessary, for instance, $\rho(s, t)$ is denoted ρ.

Substituting in Eq. (1) $\rho_k = \rho d_k$ (Eq. 2) and developing the divergence

$$\partial_t(\rho d_k) + \operatorname{div}(\rho d_k \mathbf{v}_k) = \partial_t(\rho d_k) + \partial_x(\rho d_k v_{kx}) + \partial_y(\rho d_k v_{ky}) = 0 \ ,$$

for $k = 1, 2, \dots n$. In order to introduce logarithmic derivatives, the equation is divided by ρd_k

$$\frac{\partial_t(\rho d_k)}{\rho d_k} + \frac{\partial_x(\rho d_k v_{kx})}{\rho d_k} + \frac{\partial_y(\rho d_k v_{ky})}{\rho d_k} = 0, \quad k = 1, 2, \dots n \ .$$

This equation is transformed into

$$\begin{aligned}
\partial_t \log(\rho d_k) &= -\left(v_{kx} \partial_x \log(\rho d_k) + v_{ky} \partial_y \log(\rho d_k)\right) - \left(\partial_x v_{kx} + \partial_y v_{ky}\right) \\
&= -\langle \mathbf{v}_k, \operatorname{grad} \log(\rho d_k) \rangle - \operatorname{div}(\mathbf{v}_k) \\
&= -\partial_{\mathbf{v}_k} \log(\rho d_k) - \operatorname{div}(\mathbf{v}_k) \ , \quad k = 1, 2, \dots, n \ ,
\end{aligned} \tag{3}$$

where $\langle \cdot, \cdot \rangle$ is the standard Euclidean inner product in $\mathbb{R}^2$ and known properties of derivatives of functions of several variables have been applied. Equation (3) for $k = 1, 2, \dots, n$ can be placed in an array as

$$\begin{pmatrix} \partial_t \log(\rho d_1) \\ \partial_t \log(\rho d_2) \\ \vdots \\ \partial_t \log(\rho d_n) \end{pmatrix} = \begin{pmatrix} -\partial_{\mathbf{v}_1} \log(\rho d_1) \\ -\partial_{\mathbf{v}_2} \log(\rho d_2) \\ \vdots \\ -\partial_{\mathbf{v}_n} \log(\rho d_n) \end{pmatrix} + \begin{pmatrix} -\operatorname{div}(\mathbf{v}_1) \\ -\operatorname{div}(\mathbf{v}_2) \\ \vdots \\ -\operatorname{div}(\mathbf{v}_n) \end{pmatrix} . \tag{4}$$

Logarithmic derivatives in left-hand side of Eq. (4) are transformed into a simplicial derivative by taking clr^{-1}

$$\mathcal{C}\exp\begin{pmatrix}\partial_t\log(\rho d_1)\\ \partial_t\log(\rho d_2)\\ \vdots\\ \partial_t\log(\rho d_n)\end{pmatrix} = \mathcal{C}\exp\begin{pmatrix}\partial_t\log\rho\\ \partial_t\log\rho\\ \vdots\\ \partial_t\log\rho\end{pmatrix} \oplus \mathcal{C}\exp\begin{pmatrix}\partial_t\log d_1\\ \partial_t\log d_2\\ \vdots\\ \partial_t\log d_n\end{pmatrix} = \partial_t^{\oplus}\mathbf{d}\,, \quad (5)$$

where $\mathbf{d}$ is a n-part composition obtained by the closure of the $\mathbb{R}^n$-vector which positive components are the d_k's. Note that a composition with equal components is the neutral element in $\mathbb{S}^n$ and this is the reason for cancelling the array containing the terms $\partial_t\rho$ in Eq. (5). Vectors in the right-hand side of Eq. (4) need additional definitions and properties. In the standard vector field analysis, the (spatial) gradient and divergence are useful differential linear operators. Previous Eqs. (3–4) suggest that similar concepts can be defined for space-time simplicial fields. Definitions and some properties of such operators for $d = 2$ follow.

Definition 5 (*Simplicial (spatial) gradient*) Let $\mathbf{z} : S \times T \subseteq \mathbb{R}^2 \times \mathbb{R} \to \mathbb{S}^n$ be a STSF. The simplicial (spatial) gradient is defined as a bivariate STSF, taking values in $\mathbb{S}^n \times \mathbb{S}^n$, given by

$$\mathrm{grad}^{\oplus}\mathbf{z}(s,t) = \left(\partial_x^{\oplus}\mathbf{z}(s,t)\,,\ \partial_y^{\oplus}\mathbf{z}(s,t)\right)\,.$$

Directional derivatives can be expressed as a kind of $\mathbb{R}^d$-inner product of the simplicial gradient and the direction in which the directional derivative is taken. However, the fact that simplicial derivatives are in $\mathbb{S}^n$ and spatial directions are in $\mathbb{R}^d$, introduces notational intricacies.

Proposition 5 (Simplicial gradient and directional derivatives) *Let $\mathbf{z} : S \times T \subseteq \mathbb{R}^2 \times \mathbb{R} \to \mathbb{S}^n$ be a STSF and $\mathbf{u} = (u_x, u_y) \in \mathbb{R}^2$ be a vector. Directional derivative and gradient satisfies*

$$\partial_{\mathbf{u}}^{\oplus}\mathbf{z}(s,t) = \mathbf{u} \odot \mathrm{grad}^{\oplus}\mathbf{z}(s,t)\,,$$

where $\odot$ is interpreted as a perturbation-linear combination [2]

$$\mathbf{u} \odot \mathrm{grad}^{\oplus}\mathbf{z}(s,t) = u_x \odot \partial_x^{\oplus}\mathbf{z}(s,t) \oplus u_y \odot \partial_y^{\oplus}\mathbf{z}(s,t)\,,$$

and $\mathrm{grad}^{\oplus}\mathbf{z}$ *has been decomposed in their two components $\partial_x^{\oplus}\mathbf{z}$ and $\partial_y^{\oplus}\mathbf{z}$.*

Definition 6 (*Simplicial derivative along a multiple vector field*) Let $\mathbf{z} : S \times T \subseteq \mathbb{R}^2 \times \mathbb{R} \to \mathbb{S}^n$ be a STSF with positive components z_k ($k = 1, 2, \ldots, n$). Let $\mathbf{v} = (\mathbf{v}_1, \mathbf{v}_2, \ldots, \mathbf{v}_n)$ be a multiple vector field, being $\mathbf{v}_k = (v_{kx}, v_{ky})$, $k = 1, 2, \ldots, n$. The simplicial derivative of $\mathbf{z}$ along the multiple vector field $\mathbf{v}$ is

$$\partial_{\mathbf{v}}^{\oplus}\mathbf{z}(s,t) = \mathcal{C}\exp\begin{pmatrix}\partial_{\mathbf{v}_1}\log(z_1(s,t))\\ \partial_{\mathbf{v}_2}\log(z_2(s,t))\\ \vdots\\ \partial_{\mathbf{v}_n}\log(z_n(s,t))\end{pmatrix},$$

where $\partial_{\mathbf{v}_k}\log(z_k(s,t)) = v_{kx}\partial_x\log z_k + v_{ky}\partial_y\log z_k$ is an inner product of $\mathbf{v}_k$ and $\mathrm{grad}(\log z_k)$ in $\mathbb{R}^2$.

The simplicial derivative along a multiple vector field is not linear in the simplex. A linear combination of compositions like $(\alpha_1 \odot \mathbf{z}_1) \oplus (\alpha_2 \odot \mathbf{z}_2)$ is not equal to the perturbation-linear combination of the two derivatives. However, it is linear in the simplex for linear combinations of multiple vector fields.

Definition 7 (*Simplicial divergence*) Let $\mathbf{z}_1, \mathbf{z}_2 : S \times T \subseteq \mathbb{R}^2 \times \mathbb{R} \to \mathbb{S}^n$ be two STSFs. The simplicial divergence of the pair $(\mathbf{z}_1, \mathbf{z}_2)$ is a composition in $\mathbb{S}^n$ given by

$$\begin{aligned}\mathrm{div}^{\oplus}(\mathbf{z}_1,\mathbf{z}_2) &= \partial_x^{\oplus}\mathbf{z}_1 \oplus \partial_y^{\oplus}\mathbf{z}_2\\ &= \mathrm{clr}^{-1}\left[\partial_x\mathrm{clr}(\mathbf{z}_1) + \partial_y\mathrm{clr}(\mathbf{z}_2)\right].\end{aligned}$$

In Eq. (4) the clr-inverse of vectors in right -hand side of this equation can be computed. According to definition of simplicial derivative along a multiple vector field, the first term produces

$$\mathcal{C}\exp\begin{pmatrix}-\partial_{\mathbf{v}_1}\log(\rho d_1)\\ -\partial_{\mathbf{v}_2}\log(\rho d_2)\\ \vdots\\ -\partial_{\mathbf{v}_n}\log(\rho d_n)\end{pmatrix} = \partial_{\mathbf{v}}^{\oplus}(\rho\mathbf{d}) = \partial_{\mathbf{v}}^{\oplus}(\mathbf{d}), \tag{6}$$

where $\mathbf{d} = \mathcal{C}(d_1, d_2, \ldots, d_n) \in \mathbb{S}^n$. According to definition of simplicial divergence, the second term gives

$$\mathcal{C}\exp\begin{pmatrix}-\mathrm{div}(\mathbf{v}_1)\\ -\mathrm{div}(\mathbf{v}_2)\\ \vdots\\ -\mathrm{div}(\mathbf{v}_n)\end{pmatrix} = \mathrm{div}^{\oplus}(\mathbf{w}_x,\mathbf{w}_y), \tag{7}$$

where $\mathbf{w}_x = \mathrm{clr}^{-1}(\mathbf{v}_x)$, $\mathbf{w}_y = \mathrm{clr}^{-1}(\mathbf{v}_y)$; moreover, $\mathbf{v}_x$, $\mathbf{v}_y$ are $\mathbb{R}^n$-vectors grouping the first and second components of $\mathbf{v}_k$ for $k = 1, 2, \ldots, n$, respectively. Note that $\mathrm{clr}(\mathbf{w}_x) = \mathbf{v}_x - \mathbf{v}_{x0}$, where $\mathbf{v}_{x0}$ is a constant-component vector which components are the arithmetic mean of the components of $\mathbf{v}_x$; and similarly for $\mathrm{clr}(\mathbf{w}_y)$.

Considering Eqs. (5)–(7), the compositional mass continuity equation can be written as

$$\partial_t^{\oplus}\mathbf{d} = \partial_{\mathbf{v}}^{\oplus}\mathbf{d} \oplus \mathrm{div}^{\oplus}(\mathbf{w}_x, \mathbf{w}_y)\ . \tag{8}$$

An important feature is that the overall density ρ does not appear in Eq. (8), therefore, this equation is purely compositional. Note that, in general, this is not a linear equation in the simplex due to the non-linearity of $\partial_{\mathbf{v}}^{\oplus}\mathbf{d}$ with respect to $\mathbf{d}$.

Furthermore, if for each species ρ_k is constant, then Eq. (8) reduces to

$$\mathrm{div}^{\oplus}(\mathbf{w}_x, \mathbf{w}_y) = \mathbf{n}, \quad \mathbf{n} = \mathcal{C}(1, 1, \ldots, 1)\ ,$$

quite similar to standard continuity equation for an incompressible fluid flow [8].

In some cases the simplicial fields $\mathbf{z}_1$ and $\mathbf{z}_2$ can be the two components of a simplicial gradient, that is, there is a simplicial field $\mathbf{w}$ such that $(\mathbf{z}_1, \mathbf{z}_2) = \mathrm{grad}^{\oplus}\mathbf{w} = (\partial_x^{\oplus}\mathbf{w}, \partial_y^{\oplus}\mathbf{w})$. In these cases, the bivariate simplicial field $(\mathbf{z}_1, \mathbf{z}_2)$ is said to derive from the potential composition $\mathbf{w}$, again following the ideas of the standard vector analysis. An important differential operator in this situation is the Laplacian $\Delta = \partial_x^2 + \partial_y^2$. The compositional counterpart of the Laplacian can be defined as follows:

Definition 8 (*Simplicial Laplacian*) Let $\mathbf{x}$ be a location in a plane domain R and let $\mathbf{w}$ be a STSF defined in a neighbourhood of $(\mathbf{x}, t) = (x, y, t)$. The simplicial Laplacian of $\mathbf{w}$ is a composition in $\mathbb{S}^n$ given by

$$\Delta^{\oplus}\mathbf{w} = \mathrm{div}^{\oplus}(\mathrm{grad}^{\oplus}\mathbf{w}) = \partial_x^{\oplus 2}\mathbf{w} \oplus \partial_y^{\oplus 2}\mathbf{w}\ ,$$

where the symbol $\partial^{\oplus 2}$ is the second order simplicial derivative [4].

To show the relevance of previous definitions, it is worth to state a compositional extension of the Gauss divergence theorem, here stated in for plane space.

Theorem 1 *Let $\mathbf{z}_1, \mathbf{z}_2 : S \times T \subseteq \mathbb{R}^2 \times \mathbb{R} \to \mathbb{S}^n$ be two STSF's, differentiable up to second order. Let R be a bounded and connected domain in the plain with piecewise regular and closed boundary Γ. Consider x, y as the Cartesian plain coordinates, and a piecewise regular parametrization $x = x(u)$, $y = y(u)$ of the boundary Γ. Then,*

$$\iint_R^{\oplus} (dxdy) \odot \mathrm{div}^{\oplus}(\mathbf{z}_1(x, y, t), \mathbf{z}_2(x, y, t))$$
$$= \int_{\Gamma}^{\oplus} ds \odot \left(\partial_{\mathbf{n}_x}\mathbf{z}_1(x(u), y(u), t) \oplus \partial_{\mathbf{n}_y}\mathbf{z}_2(x(u), y(u), t)\right)\ ,$$

where $\mathbf{n}_x = -y'(u)/\sqrt{x'(u)^2 + y'(u)^2}$, $\mathbf{n}_y = x'(u)/\sqrt{x'(u)^2 + y'(u)^2}$ are, respectively, the vector fields of the first and second components of the normal direction to the boundary Γ; and $ds = \sqrt{x'(u)^2 + y'(u)^2}\, du$.

4 Conclusions

The study of space-time simplicial fields reveals some interesting aspects and properties for applications in problems related with space-time evolutionary compositions. The present contribution is not complete and needs to be developed further.

It is possible to define, in a natural way, simplicial differential operators, similar to standard equations appearing in fluid mechanics and vector fields in general. The continuity equation of mass in fluid mechanics has been studied in some detail as a motivation to introduce some definitions. However, one of the needed simplicial differential operators, is not linear in the simplex, thus introducing features which are not dealt with in the standard formulation. It seems possible and useful to study the simplicial version of some important equations in Fluid Mechanics and other parts of Physics, such as advection–diffusion, Navier–Stokes and others in their compositional formulation.

Acknowledgments This research has been supported by the *Spanish Ministry of Education, Culture ans Sport* under project 'CODA-RETOS / TRANSCODA' (Ref. MTM2015-65016-C2-1-R and MTM2015-65016-C2-2-R) and supported by the *Agència de Gestió d'Ajuts Universitaris i de Recerca (AGAUR)* of the *Generalitat de Catalunya (GENCAT)* under the project "Compositional and Spatial Analysis" (COSDA, Ref: 2014SGR551; 2014–2016).

References

1. Aitchison, J.: The Statistical Analysis of Compositional Data (Reprinted in 2003 by The Blackburn Press), p. 416. Chapman & Hall Ltd., London (UK) (1986)
2. Egozcue, J.J., Barceló-Vidal, C., Martín-Fernández, J.A., Jarauta-Bragulat, E., Díaz-Barrero, J.L., Mateu-Figueras, G.: Elements of simplicial linear algebra and geometry. See Pawlowsky-Glahn and Buccianti (2011)
3. Egozcue, J.J., Jarauta-Bragulat, E.: Differential models for evolutionary compositions. Math. Geosci. **46**(4), 381–410 (2014)
4. Egozcue, J.J., Jarauta-Bragulat, E., Díaz-Barrero, J.L.: Calculus of simplex-valued functions. See Pawlowsky-Glahn and Buccianti (2011)
5. Egozcue, J.J., Pawlowsky-Glahn, V., Mateu-Figueras, G., Barceló-Vidal, C.: Isometric logratio transformations for compositional data analysis. Math. Geol. **35**(3), 279–300 (2003)
6. Landau, L.D., Lifshitz, E.M.: Fluid mecahnics. In: Course of Theoretical Physics, vol. 6, 2nd edn, p. 531. Pergamon Press, New York (1987)
7. Pawlowsky-Glahn, V., Buccianti, A. (eds.): Compositional Data Analysis: Theory to Applications. Wiley & Sons (2011)
8. White, F.M.: Viscous Fluid Flow. Mechanical Engineering, p. 614. McGraw Hill International Editions (1991)

A Regression Model for Compositional Data Based on the Shifted-Dirichlet Distribution

G.S. Monti, G. Mateu-Figueras, V. Pawlowsky-Glahn and J.J. Egozcue

Abstract Using an approach based on the Aitchison geometry of the simplex, a Shifted-Dirichlet covariate model is obtained. Allowing the parameters to change linearly with a set of covariates, their effects on the relative contributions of different components in a composition are assessed. An application of this model to sedimentary petrography is given.

Keywords Dirichlet regression · Simplicial regression · Model selection

1 Introduction

Compositional data are vectors of parts of some whole which carry relative information. They are frequently represented as proportions or percentages, which are subject to a constant sum, κ, i.e., $\kappa = 1$ or $\kappa = 100$. Their sample space is then represented by the simplex, denoted by

G.S. Monti (✉)
Department of Economics, Management and Statistics, University of Milano-Bicocca, Milano, Italy
e-mail: gianna.monti@unimib.it

G. Mateu-Figueras · V. Pawlowsky-Glahn
Department of Computer Science, Applied Mathematics and Statistics, University of Girona, Girona, Spain
e-mail: gloria.mateu@udg.edu

V. Pawlowsky-Glahn
e-mail: vera.pawlowsky@udg.edu

J.J. Egozcue
Department of Civil and Environmental Engineering, Technical University of Catalonia, Barcelona, Spain
e-mail: juan.jose.egozcue@upc.edu

J.A. Martín-Fernández and S. Thió-Henestrosa (eds.), *Compositional Data Analysis*, Springer Proceedings in Mathematics & Statistics 187,
DOI 10.1007/978-3-319-44811-4_9

$$\mathscr{S}^D = \{\mathbf{x} = (x_1, \ldots, x_D),\ x_i > 0,\ \sum_{i=1}^{D} x_i = \kappa\}.$$

Compositional data occur in many applied fields: from geology and biology to election forecast, from medicine and psychology to economic studies.

We recall briefly the essential elements of simplicial algebra, as it will be used later. For any vector of D strictly positive real components,

$$\mathbf{z} = (z_1, \ldots, z_D) \in \mathbb{R}_+^D \quad z_i > 0,\ \text{for all } i = 1, \ldots, D,$$

the *closure* operation of $\mathbf{z}$ is defined as

$$\mathscr{C}(\mathbf{z}) = \left(\frac{\kappa\ z_1}{\sum_{i=1}^{D} z_i}, \ldots, \frac{\kappa\ z_D}{\sum_{i=1}^{D} z_i} \right) \in \mathscr{S}^D. \tag{1}$$

where κ is the sum of the components, i.e., the constraint.

The two basic operations required for a vector space structure of the simplex are *perturbation*: given two compositions $\mathbf{x}$ and $\mathbf{y} \in \mathscr{S}^D$,

$$\mathbf{x} \oplus \mathbf{y} = \mathscr{C}\left(x_1 y_1, \ldots, x_D y_D\right), \tag{2}$$

and *powering*: given a composition $\mathbf{x} \in \mathscr{S}^D$ and a scalar $\alpha \in \mathbb{R}$,

$$\alpha \odot \mathbf{x} = \mathscr{C}\left(x_1^\alpha, \ldots, x_D^\alpha\right). \tag{3}$$

Furthermore, an *inner product* $\langle \cdot, \cdot \rangle_a$ is defined as

$$\langle \mathbf{x}, \mathbf{y} \rangle_a = \sum_{i=1}^{D} \ln \frac{x_i}{\mathrm{g_m}(\mathbf{x})} \ln \frac{y_i}{\mathrm{g_m}(\mathbf{y})}, \tag{4}$$

where $\mathrm{g_m}(\mathbf{x})$ denotes the geometric mean of the components of $\mathbf{x}$ [4, 22]. As shown in Pawlowsky-Glahn and Egozcue [22] the simplex $(\mathscr{S}^D, \oplus, \odot, \langle \cdot, \cdot \rangle_a)$ has a $(D-1)$-dimensional real Euclidean vector space structure called *simplicial* or *Aitchison geometry*.

Let $(\mathbf{e}_1, \mathbf{e}_2, \ldots, \mathbf{e}_{D-1})$ be an orthonormal basis of the simplex and consider the $(D-1) \times D$ matrix $\boldsymbol{\Psi}$ which rows are $\Psi_i = \mathrm{clr}(\mathbf{e}_i)$, $(i = 1, \ldots, D-1)$. Note that clr is the centered log-ratio transformation, a function from $\mathscr{S}^D$ to $\mathbb{R}^D$ defined as

$$\mathrm{clr}(\mathbf{x}) = \left(\log \frac{x_1}{\mathrm{g_m}(\mathbf{x})}, \ldots, \log \frac{x_D}{\mathrm{g_m}(\mathbf{x})} \right),$$

where $g_m(\mathbf{x})$ is the geometric mean of the D components of $\mathbf{x}$. The $\boldsymbol{\Psi}$ matrix is called *contrast matrix* associated with the orthonormal basis $(\mathbf{e}_1, \ldots, \mathbf{e}_{D-1})$. Each row is called a (log)contrast.

The *isometric log-ratio transformation*, ilr for short, of $\mathbf{x}$ is the function ilr : $\mathcal{S}^D \to \mathbb{R}^{D-1}$, which assigns coordinates $\mathbf{x}^\star$, with respect to the given basis, to the composition $\mathbf{x}$. The vector $\mathbf{x}^\star$ contains the $D-1$ ilr-coordinates of $\mathbf{x}$ in a Cartesian coordinate system. The inverse of the ilr-transformation is denoted by ilr^{-1}. The function ilr is an isometry of vector spaces. The ilr transformation is computed as a simple matrix product:

$$\mathbf{x}^\star = \text{ilr}(\mathbf{x}) = \ln(\mathbf{x})\boldsymbol{\Psi}' .$$

Inversion of ilr, i.e., recovering the composition from its coordinates, is given by

$$\mathbf{x} = \text{ilr}^{-1}(\mathbf{x}^\star) = \mathscr{C}\left(\exp(\boldsymbol{\Psi}\,\mathbf{x}^\star)\right) .$$

Given an orthonormal basis of the simplex, any composition $\mathbf{x} \in \mathcal{S}^D$ can be expressed as a linear combination,

$$\begin{aligned}\mathbf{x} &= (x_1^\star \odot \mathbf{e}_1) \oplus (x_2^\star \odot \mathbf{e}_2) \oplus \cdots \oplus (x_{D-1}^\star \odot \mathbf{e}_{D-1}) \\ &= \bigoplus_{i=1}^{D-1} (x_i^\star \odot \mathbf{e}_i),\end{aligned}$$

where the symbol $\bigoplus$ represents repeated perturbation. The coefficients of the linear combination, for a fixed basis, are uniquely determined, given that in a Euclidean space any point can always be represented in a unique way by its coordinates with respect to an orthonormal basis. Once an orthonormal basis has been chosen, all standard statistical methods can be applied to coordinates and transferred to the simplex preserving their properties [15].

A natural measure on $\mathcal{S}^D$, called *Aitchison measure*, can be defined using orthonormal coordinates [21, 23], that is, the Aitchison measure of a subset on the simplex is the Lebesgue measure of the subset in the space of orthonormal coordinates. This measure is compatible with the *Aitchison geometry* and is absolutely continuous with respect to the Lebesgue measure on the D-dimensional real space. The relationship between them is $\sqrt{D}x_1x_2\ldots x_D$. The change of the reference measure has some important implications, for example to compute the expected value (see [16] for an in-depth discussion).

Historically, there are essentially two different approaches to regression models which relate a compositional response variable with a system of covariates: Simplicial regression and Dirichlet regression. The former is based on the Aitchison's theoretical result that if a compositional vector follows an additive logistic normal distribution, the log-ratio transformed vector will follow a normal distribution [2, 3, 8]; the latter follows the *stay-in-the-simplex* approach. It assumes that the response variable follows a Dirichlet distribution whose parameters are a log-linear function of a set of covariates [6, 12, 14, 17]. Other solutions, present in the literature but less used, involve models based on the generalized Liouville distribution [25].

The Scaled-Dirichlet distribution is an extension of the Dirichlet one. Given that we work here with the Aitchison geometry of the simplex, and that within this framework it is a perturbation of a standard Dirichlet [18], we will refer hereafter to it as the Shifted-Dirichlet distribution. The reason to change this terminology is twofold. On the one hand, working within the Aitchison geometry implies a change of the reference measure; on the other hand, scaling in this geometry is achieved using a power transformation, which allows another extension already studied in Monti et al. [19]. In summary, the name of the distribution indicates the sample space of the corresponding random vector and its structure. For the Scaled-Dirichlet distribution this is the simplex as a subset of real space with the induced Euclidean geometry, while for the Shifted-Dirichlet distribution it is the simplex as a Euclidean space endowed with the Aitchison geometry. Although in the first case the Lebesgue reference measure is used, and in the second the Aitchison measure, the probability assigned to any measurable subset of the simplex is the same.

The Shifted-Dirichlet covariate model is an extension of the Dirichlet one, based on the algebraic geometric structure of the simplex. The assumption is that $\mathbf{x} = (x_1, \ldots, x_D)$ is a compositional response vector, with D components having a Shifted-Dirichlet distribution, in which the parameters $\boldsymbol{\alpha}$ are allowed to change with a set of covariates.

The paper is structured as follows. Section 2 defines the two existing approaches: Simplicial regression and Dirichlet regression. Section 3 gives a brief overview of the Shifted-Dirichlet distribution and describes the Shifted-Dirichlet covariate model, dealing with the issue of parameter estimation. Section 4 presents an example of application of the proposed regression model to sedimentary petrography, in particular bulk petrography and heavy-mineral data of Pleistocene sands (Regione Lombardia cores; Po Plain); this dataset is described in Garzanti et al. [11].

2 Regression Models for Compositional Response Variable

2.1 Simplicial Regression

Linear regression with compositional response variable can be stated as follows. A compositional sample of n independent observations, denoted by $\mathbf{x}_1, \ldots, \mathbf{x}_n$, is available. Each data point, $\mathbf{x}_j$, $(j = 1, \ldots, n)$ is associated with one or more external variables or covariates grouped in the vector $\mathbf{s}_j = (s_{j0}, s_{j1}, \ldots, s_{jm}, \ldots, s_{jp})$, where $s_{j0} = 1$ by convention.

The basic idea of Simplicial regression [8] relies on the principle of working on coordinates: once a basis is chosen, the associated ilr coordinates are computed and the classical regression of the ilr coordinates on the covariates is performed. Through the inverse ilr-transformation, the back-transformed coefficient vectors, as well as predictions and confidence intervals are obtained.

The general model can be expressed as

$$\hat{\mathbf{x}}(\mathbf{s}) = (s_0 \odot \boldsymbol{\delta}_0) \oplus (s_1 \odot \boldsymbol{\delta}_1) \oplus \cdots \oplus \left(s_p \odot \boldsymbol{\delta}_p\right) = \bigoplus_{m=0}^{p} s_m \odot \boldsymbol{\delta}_m \,. \tag{5}$$

Note that there are $p + 1$ coefficient vectors δ_m, as many as covariates, and that they are vectors with $(D - 1)$ components, as many as coordinates. The goal of estimating the coefficients $\boldsymbol{\delta}$ of a curve or surface in $\mathcal{S}^D$ is solved by translating it into a $(D - 1)$ least square problem, i.e., for each coordinate

$$\hat{\mathbf{x}}_i^{\star}(\mathbf{s}) = \delta_{0i}^{\star} s_0 + \delta_{1i}^{\star} s_1 + \cdots + \delta_{pi}^{\star} s_p, \quad i = 1, \ldots, D-1, \tag{6}$$

where $\boldsymbol{\delta}_m^{\star} = \left(\delta_{m1}^{\star}, \ldots, \delta_{m,D-1}^{\star}\right)$ is the coordinate vector associated with $\boldsymbol{\delta}_m$. In the case of simple regression $m = 1$ and $\mathbf{s} = s$, which is a straight-line in the simplex.

2.2 *Dirichlet Regression*

The Dirichlet distribution is one of the well known probability models suitable for random compositions. A random vector $\mathbf{X} = (X_1, \ldots, X_D) \in \mathcal{S}^D$ has a Dirichlet distribution, indicated by $\mathbf{X} \sim \mathcal{D}^D(\boldsymbol{\alpha})$, with $\boldsymbol{\alpha} = (\alpha_1, \ldots, \alpha_D) \in \mathbb{R}_+^D$, when its density function (with respect to the Aitchison measure) is

$$f(\mathbf{x}; \boldsymbol{\alpha}) = \frac{\sqrt{D}\,\Gamma(\alpha_+)}{\prod_{i=1}^{D} \Gamma(\alpha_i)} \prod_{i=1}^{D} x_i^{\alpha_i}, \tag{7}$$

where $\alpha_+ = \sum_{i=1}^{D} \alpha_i$, and Γ denotes the gamma function [18]. The Dirichlet distribution has D parameters α_i, which are assumed to be positive. Note that the density (7) is obtained by changing the measure to a Dirichlet density with respect to the Lebesgue measure.

In the Dirichlet regression model the α_i parameters are reparameterized in terms of explanatory variables and coefficients through an exponential function as described in Eq. (12). The log-likelihood of the reparameterized Dirichlet distribution can be optimized via an iterative method such as the Newton–Raphson algorithm.

When the variable of interest is continuous and restricted to the unit interval (0, 1), i.e., when $D = 2$, the Dirichlet regression is called Beta regression [10].

3 Shifted-Dirichlet Covariate Model

3.1 Shifted-Dirichlet Distribution

One of the generalizations of the Dirichlet distribution is the Shifted-Dirichlet distribution. A random vector $\mathbf{X} \in \mathcal{S}^D$ has a Shifted-Dirichlet distribution with parameters $\boldsymbol{\alpha}$ and $\boldsymbol{\beta} = (\beta_1, \ldots, \beta_D) \in \mathcal{S}^D$ if its density function is

$$f(\mathbf{x}, ; \boldsymbol{\alpha}, \boldsymbol{\beta}) = \frac{\Gamma(\alpha_+)\sqrt{D}}{\prod_{i=1}^D \Gamma(\alpha_i)} \frac{\prod_{i=1}^D \left(\frac{x_i}{\beta_i}\right)^{\alpha_i}}{\left(\sum_{i=1}^D \frac{x_i}{\beta_i}\right)^{\alpha_+}}, \tag{8}$$

The density (8) is expressed with respect to the Aitchison probability measure [21]. See Monti et al. [18] for a detailed discussion about the reasons and implications to use the Aitchison measure. This distribution will be denoted by $\mathbf{X} \sim \mathcal{SD}^D(\boldsymbol{\alpha}, \boldsymbol{\beta})$.

The number of parameters of this model is $2D - 1$, since $\boldsymbol{\beta} \in \mathcal{S}^D$. The Shifted-Dirichlet distribution can be obtained by normalizing a vector of D independent, scaled (in the Euclidean geometry of real space), gamma r.v.s $W_i \sim Ga(\alpha_i, \beta_i)$, $i = 1, 2, \ldots, D$; i.e., if $\mathbf{X} = \mathcal{C}(\mathbf{W})$, with $\mathbf{W} = (W_1, \ldots, W_D) \in \mathbb{R}_+^D$, then $\mathbf{X} \sim \mathcal{SD}^D(\boldsymbol{\alpha}, \boldsymbol{\beta})$ [18]. For this reason, in the literature, when working with the Lebesgue reference measure, the distribution is called Scaled-Dirichlet. This distribution can also be obtained as a perturbed random composition with a Dirichlet density. Recall that perturbation is, in the Aitchison geometry of the simplex, a shift. Therefore, here it is called Shifted-Dirichlet distribution, understanding that the density is expressed with respect to the Aitchison measure.

Indeed, let $\widetilde{\mathbf{X}} \sim \mathcal{D}^D(\boldsymbol{\alpha})$ be a random composition defined in $\mathcal{S}^D$, and let $\boldsymbol{\beta} \in \mathcal{S}^D$ be a composition. The random composition $\mathbf{X} = \ominus\boldsymbol{\beta} \oplus \widetilde{\mathbf{X}}$ has a $\mathcal{SD}^D(\boldsymbol{\alpha}, \boldsymbol{\beta})$ distribution (note that $\ominus$ is the inverse operation of $\oplus$). Observe that using the Aitchison measure and geometry, $\boldsymbol{\beta}$ can be interpreted as a parameter of location, instead of as a measure of scale. The expected value of $\mathbf{X} \sim \mathcal{SD}^D(\boldsymbol{\alpha}, \boldsymbol{\beta})$ with respect to the Aitchison measure is

$$\mathrm{E}_a(\mathbf{X}) = \ominus\boldsymbol{\beta} \oplus \mathrm{E}_a(\widetilde{\mathbf{X}}), \tag{9}$$

where $\mathrm{E}_a(\widetilde{\mathbf{X}})$ is the expected value of a Dirichlet composition with respect to the Aitchison measure

$$\mathrm{E}_a(\widetilde{\mathbf{X}}) = \mathcal{C}(e^{\psi(\alpha_1)}, \ldots, e^{\psi(\alpha_D)}), \tag{10}$$

with ψ the digamma function. The metric variance of $\mathbf{X}$ coincides with the metric variance of a Dirichlet composition, because this measure of dispersion is invariant under perturbation

$$\mathrm{Mvar}(\mathbf{X}) = \frac{D-1}{D}(\psi'(\alpha_1), \ldots, \psi'(\alpha_D)), \tag{11}$$

with ψ' the trigamma function [1].

3.2 Shifted-Dirichlet Regression

Given a sample of n independent compositional observations $(\mathbf{x}_1, \ldots, \mathbf{x}_n)$, we hypothesize that each observation $\mathbf{x}_j$ follows a conditional Shifted-Dirichlet distribution, given a set of covariates. Polynomial regression on a covariate s is included as a particular case taking $s_{jm} = s_j^m$.

In order to incorporate the covariate effects into the model [6, 14], we reparameterize each parameter α_i of the density written in Eq. (8) in terms of covariates and regression coefficients via the following log-linear model

$$\alpha_{ij} = \alpha_{ij}(\mathbf{s}_j) = \exp\left\{\sum_{m=0}^{p} \delta_{im} s_{jm}\right\}, \qquad (12)$$

where $\mathbf{s}_j$ is the covariate vector recorded on the j-th observed composition ($j = 1, \ldots, n$), and δ_{im} are the coefficients for the m-th covariate. The parameter δ_{im} theoretically can vary by component, and the covariates may or may not be the same set of explanatory variables for each α_{ij}. We augment each vector $\mathbf{s}_j$ with 1 as first position for notation simplicity. Thus, given a sample of independent compositional observations of size n, $\mathbf{x}_1, \ldots, \mathbf{x}_j, \ldots, \mathbf{x}_n$ the log-likelihood function for the reparameterized Shifted-Dirichlet, given the covariates $\mathbf{s}$ and ignoring the constant part that does not involve the parameters, is equal to

$$\begin{aligned} l(\boldsymbol{\beta}, \boldsymbol{\delta}|\mathbf{x}, \mathbf{s}) = \sum_{j=1}^{n} \Bigg\{ & \log \Gamma\left(\sum_{i=1}^{D} \exp\left\{\sum_{m=0}^{p} \delta_{im} s_{jm}\right\}\right) - \sum_{i=1}^{D} \log \Gamma\left(\exp\left\{\sum_{m=0}^{p} \delta_{im} s_{jm}\right\}\right) \\ & - \sum_{i=1}^{D} \exp\left\{\sum_{m=0}^{p} \delta_{im} s_{jm}\right\} \log \beta_i + \sum_{i=1}^{D} \exp\left\{\sum_{m=0}^{p} \delta_{im} s_{jm}\right\} \log x_{ij} \\ & - \left(\sum_{i=1}^{D} \exp\left\{\sum_{m=0}^{p} \delta_{im} s_{jm}\right\}\right) \log\left(\sum_{i=1}^{D} \frac{x_i}{\beta_i}\right) \Bigg\}. \end{aligned} \qquad (13)$$

Equation (13) can be estimated using the maximum likelihood method via some optimization algorithm, e.g., the Newton–Raphson algorithm. The choice of the starting values for the algorithm is of fundamental importance to get fast convergence.

For the Dirichlet regression in Hijazi and Jernigan [14] a method based on resampling from the original data is proposed; for each resample a Dirichlet model with constant parameters is fitted and the mean of the corresponding covariates is computed. After that, D models of the form $\sum_{m=0}^{p} \delta_{im} s_{jm}$ are fitted by least squares. The fitted coefficients $\hat{\delta}_{im}$ are used as starting values. For the Shifted-Dirichlet covariate model we have followed the same principle; as starting point for the vector $\boldsymbol{\beta}$ we have chosen the closed geometric mean of the components of $\mathbf{x}$ given by

$$\mathbf{g}(\mathbf{x}) = \mathscr{C}\left(\left(\prod_{j=1}^{n} x_{1j}\right)^{1/n}, \ldots, \left(\prod_{j=1}^{n} x_{Dj}\right)^{1/n}\right), \tag{14}$$

Model selection is performed by testing

$$H_0:\ \delta_{im} = 0\ , \tag{15}$$

for some pair $(i, m), i = 1, \ldots, D$ and $m = 1, \ldots, p$. For it, the traditional likelihood ratio test is implemented.

4 Example from Sedimentology

4.1 Data Description

In Garzanti et al. [11] the authors studied the paleogeographic and paleodrainage changes during Pleistocene glaciations of Po Plain by compositional signatures of Pleistocene sands. In particular we consider here Cilavegna and Ghedi cores of Regione Lombardia, with 18 and 19 compositional observations, respectively. In this section we compare the above-mentioned approaches to regression with a composition as dependent variable. The goal is to model the effect of the depth covariate on compositional signatures of Pleistocene sands taking the fact into account that the cores may have separate effects on the response. The three compositional parts are: Q (= quartz), F (= feldspar), and L (= lithic grains) represented in the usual ternary diagram in Fig. 1.

4.2 Estimated Models Comparison

For each of the three components we have fitted a regression model considering the depth as covariate, including in the model one dummy variable representing the core provenance (0 for Cilavegna and 1 for Ghedi), as well as the interaction term. The categorical covariate core has been included to account for variation in proportions that is a function of a group-specific factor. The Dirichlet and Shifted-Dirichlet covariate models are expressed with respect to the Aitchison measure.

Tables 1 and 2 summarize the estimated regression coefficients, together with results from the inference for the Dirichlet and Shifted-Dirichlet covariate model, respectively.

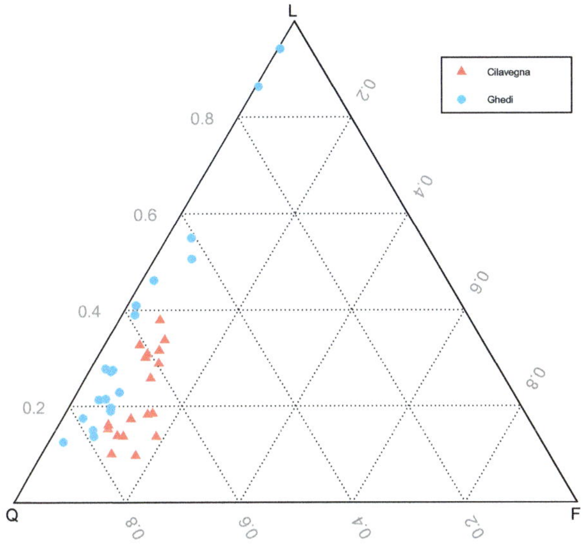

Fig. 1 Quaternary Po Plain sediments: ternary plot. Points are distinguished by core

Table 1 Regression output of the Dirichlet covariate model for Quaternary Po Plain sediments

Regressors	Coefficient	S.E.	z value	Pr(> |z|)
	δ-coefficients for variable Q (= quartz)			
(Intercept)	3.8695	0.8536	4.5329	0.0000
Depth	0.0027	0.0063	0.4304	0.6669
Core	−4.6905	0.5509	−8.5137	0.000
Depth ∗ core	0.0243	0.0038	6.4577	0.0000
	δ-coefficients for variable F (= feldspar)			
(Intercept)	2.2494	0.7975	2.8205	0.0048
Depth	0.0015	0.0058	0.2495	0.8030
Core	−4.3685	0.6012	−7.2662	0.0000
Depth ∗ core	0.0190	0.0039	4.9164	0.0000
	δ-coefficients for variable L (= lithic grains)			
(Intercept)	2.0708	0.7773	2.6641	0.0077
Depth	0.0091	0.0058	1.5794	0.1143
Core	−2.3527	0.3583	−6.5655	0.0000
Depth ∗ core	0.0076	0.0021	3.5803	0.0003

Table 2 Regression output of the Shifted-Dirichlet covariate model for Quaternary Po Plain sediments

	δ-coefficients for variable Q (= quartz)			
Regressors	Coefficient	S.E.	z value	Pr(> |z|)
(Intercept)	3.1275	0.7534	4.1510	0.0000
Depth	0.0035	0.0053	0.6649	0.5061
Core	−3.5720	0.8824	−4.0481	0.0001
Depth ∗ core	0.0171	0.0069	2.4658	0.0137
	δ-coefficients for variable F (= feldspar)			
Regressors	Coefficient	S.E.	z value	Pr(> |z|)
(Intercept)	3.2542	0.7161	4.5442	0.0000
Depth	0.0023	0.0050	0.4549	0.6492
Core	−4.5289	0.7993	−5.6664	0.0000
Depth ∗ core	0.0187	0.0064	2.9332	0.0034
	δ-coefficients for variable L (= lithic grains)			
Regressors	Coefficient	S.E.	z value	Pr(> |z|)
(Intercept)	1.4002	0.6546	2.1390	0.0324
Depth	0.0098	0.0049	2.0100	0.0444
Core	−1.2044	0.8589	−1.4022	0.1608
Depth ∗ core	0.0004	0.0068	0.0516	0.9589
	β-coefficients			
Q	0.4693	0.1362	3.4466	0.0006
F	0.0789	0.0397	1.9895	0.0467

Table 3 Model fit statistics for the nested Shifted-Dirichlet covariate models, where $\Delta G^2 = -2 \log L(\text{reduced model} - \text{current model})$

Criterion	Intercept only	Depth covariate	Full model
AIC	210.1447	187.4468	122.5
BIC	218.1993	200.3341	145.053
logL	−100.0724	−85.7233	−47.25
df	5	8	14
ΔG^2		28.697	76.947
Pr > ChiSq		<0.0001	<0.0001

Table 4 Regression output for the first and second coordinate of Simplicial regression for the Quaternary Po Plain sediments

Regressors	Coefficient	S.E.	z value	Pr(> $\|z\|$)
	$x_1^\star$ *coordinate*			
(Intercept)	−1.1867	0.1264	−9.3862	0.0000
Depth	−0.0008	0.0011	−0.7209	0.4760
Core	−1.0654	0.1886	−5.6487	0.0000
Depth $*$ core	0.0032	0.0015	2.1193	0.0417
	$x_2^\star$ *coordinate*			
(Intercept)	−0.8479	0.2903	−2.9212	0.0062
Depth	0.0061	0.0024	2.5012	0.0175
Core	2.7908	0.4330	6.4454	0.0000
Depth $*$ core	−0.0174	0.0035	−4.9827	0.0000

The p-value of the likelihood ratio tests to compare the intercept-only model (e.g., no predictors) with the fitted Shifted Dirichlet covariate model is essentially zero (<0.0001), which provides evidence against the reduced model in favor of the current model, as well as the model with only depth as covariate (see Table 3).

In Table 2 it can be seen that the z-values for the first two components are highly significant, implying that the use of the dummy variable is important; the models appear to have a definite nonzero slope. This consideration is confirmed by the fitted regression lines reported in Fig. 2.

Akaike information criterion (AIC) and Bayesian information criterion (BIC) are usually used to compare adequacy of models of the same family, the preferred model is the one with the minimum AIC value or BIC value. For the Dirichlet regression we obtained that $AIC = 131.2837$ and $BIC = 150.6148$, while, for the Shifted-Dirichlet regression the two measures are $AIC = 122.45$ and $BIC = 145.053$ (see Full model in Table 3). Therefore we can conclude that the improvement of the model compensates the additional parameters in the Shifted-Dirichlet model.

In order to apply the Simplicial regression, ilr coordinates of the Quaternary Po Plain sediment dataset are computed (Table 4). The canonical basis in the clr plane was used as ilr transform, so that, the two coordinates or balances are expressed as:

$$x_1^\star = \frac{1}{\sqrt{2}} \log \frac{x_2}{x_1}\,, \quad x_2^\star = \frac{1}{\sqrt{6}} \log \frac{x_3^2}{x_1\, x_2}\,. \tag{16}$$

Predictions of the three coordinates can be back-transformed with the inverse ilr, to obtain a prediction of the proportions themselves.

In order to assess the adequacy of the different regression approaches, we examine some goodness of fit measures. One suitable determination coefficient for the regression model to evaluate the proportion of explained variation in the compositions by the covariate is connected with the total variability [2, 13], based on the variation

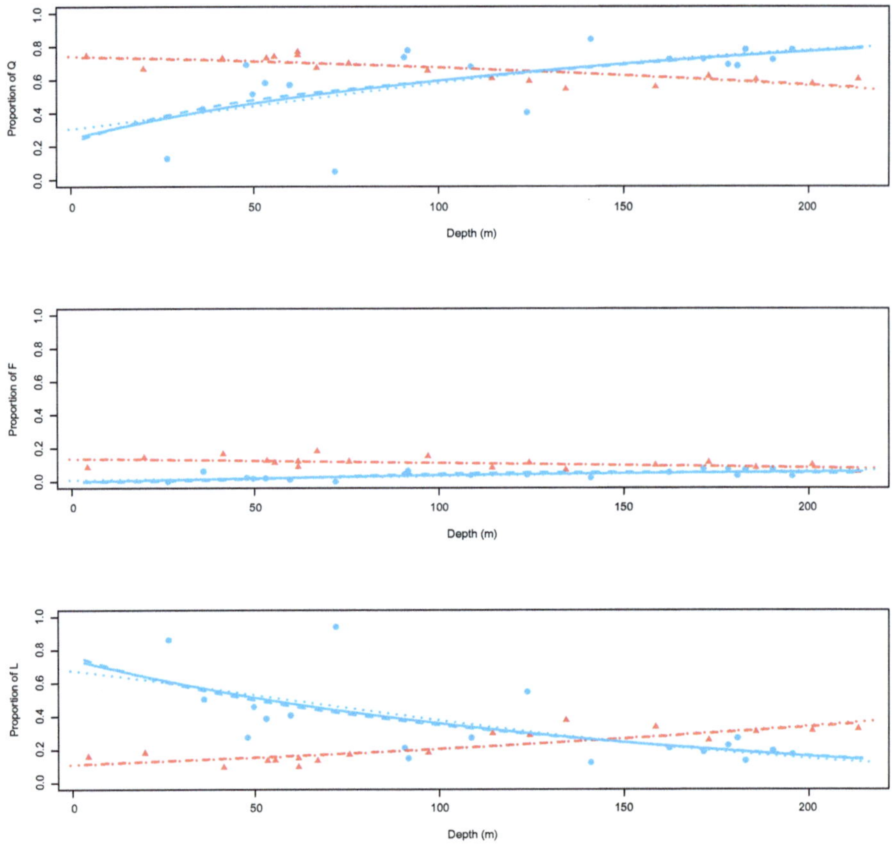

Fig. 2 Observed and fitted compositions for Quaternary Po Plain sediments using the three models for each level of the core variable (Shifted-Dirichlet covariate model: *solid line*, Dirichlet covariate model: *dashed line*; Simplicial regression: *dotted line*). *Red colors* refers to Cilavegna core data and *blue color* refers to Ghedi core data

matrix of the transformed log-ratio data,

$$\mathbf{T}(\mathbf{x}) = [t_{ir}] = \left[\operatorname{var}\left(\ln \frac{x_i}{x_r}\right)\right] \quad i, r = 1, \ldots, D. \tag{17}$$

Each element t_{ir} is the usual variance of the log ratio of parts i and r. Aitchison's total variability measure totvar($\mathbf{x}$), a measure of global dispersion of a compositional sample, is defined as

$$\operatorname{totvar}(\mathbf{x}) = \frac{1}{2D} \sum_{i,r} \operatorname{var}\left(\ln \frac{x_i}{x_r}\right) = \frac{1}{2D} \sum_{i,r} t_{ir} \,, \tag{18}$$

The determination coefficient R_T^2 is defined as

$$R_T^2 = \frac{\text{totvar}(\hat{\mathbf{x}})}{\text{totvar}(\mathbf{x})}; \tag{19}$$

it compares the total variability of the observed with the fitted data.

Table 5 Goodness of fit measures for the three different regression models

	R_T^2	R_A^2	KL-div
Dirichlet regression	0.6914	0.5624	1.6472
Shifted-Dirichlet regression	0.5733	0.5965	1.6209
Simplicial regression	0.5907	0.5907	1.657

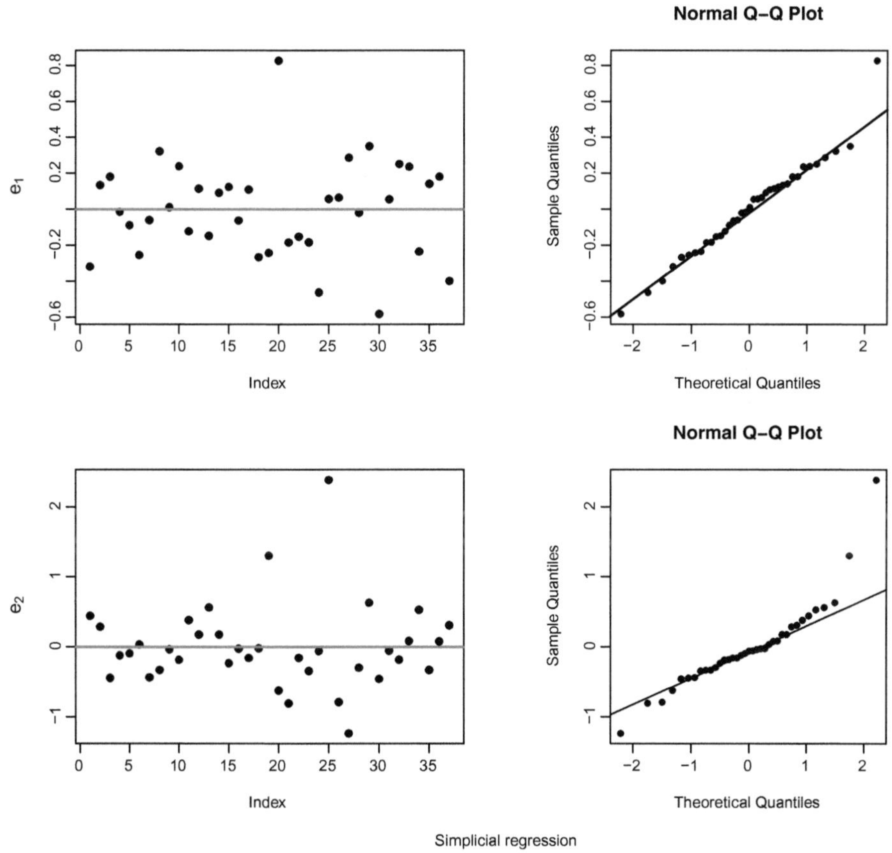

Fig. 3 In the *left column* coordinate residual plots associated to the estimated simplicial regression. In the *right column* Q–Q plots for residuals of the corresponding regression models are displayed

Moreover, the Aitchison distance of any two compositions $\mathbf{x}$ and $\mathbf{y} \in \mathcal{S}^D$ is defined as

$$d_a(\mathbf{x}, \mathbf{y}) = \sqrt{\frac{1}{2D} \sum_{i=1}^{D} \sum_{r=1}^{D} \left(\ln \frac{x_i}{x_r} - \ln \frac{y_i}{y_r} \right)^2}. \tag{20}$$

Similarly to the standard least squares regression analysis, the compositional total sum of squares (CSST) and the compositional sum of squared residuals (CSSE) are given by $CSST = \sum_{j=1}^{n} d_a^2(\mathbf{x}_j, \mathrm{g_m}(\mathbf{x}))$ and $CSSE = \sum_{j=1}^{n} d_a^2(\mathbf{x}_j, \hat{\mathbf{x}}_j)$. In this way another R^2-measure based on the compositional sum of squares [13] is

$$R_A^2 = 1 - \frac{CSSE}{CSST}. \tag{21}$$

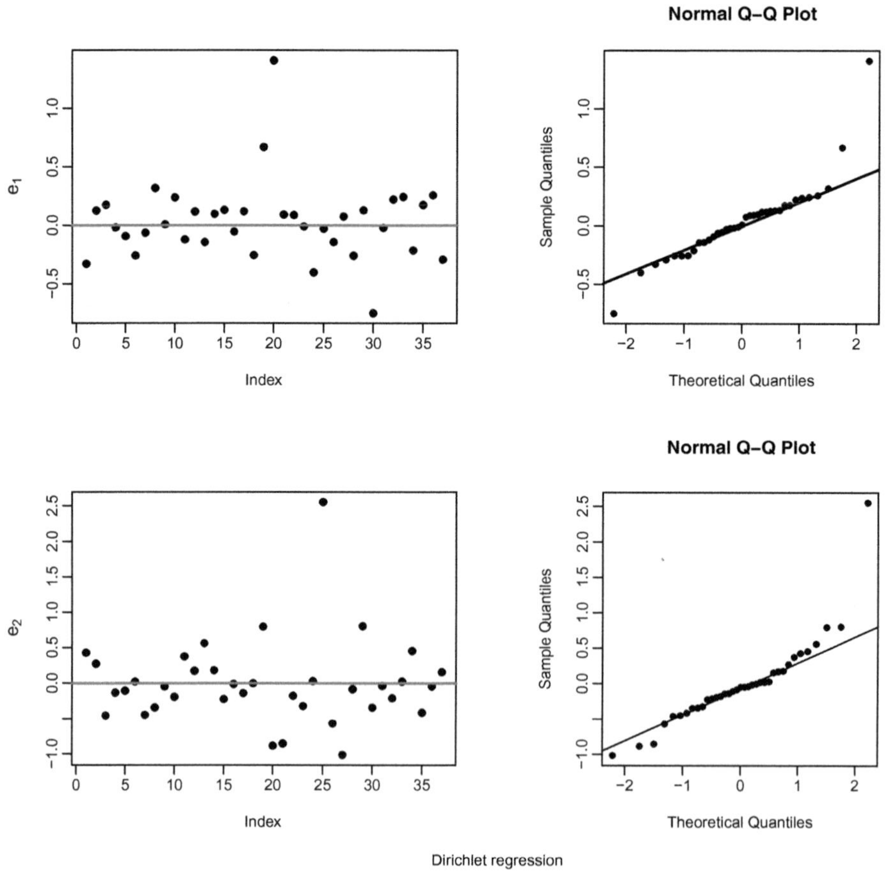

Fig. 4 In the *left column* coordinate residual plots associated to the estimated Dirichlet regression. In the *right column* Q–Q plots for residuals of the corresponding regression models are displayed

In Table 5, KL-div refers to the Kullback–Leibler divergence calculated as

$$\sum_{j=1}^{n}\sum_{i=1}^{3} x_{ji} \log \frac{x_{ji}}{\hat{x}_{ji}} .$$

The measures of goodness of fit reported in Table 5 show a good performance of the Shifted-Dirichlet model with respect to the other two models. The coefficient of determination based on the Aitchison norm shows that 60 % of the total variability is captured by the Shifted-Dirichlet regression model.

In order to check for absence of trends in central tendency and in variability, diagnostic plots are useful. We have expressed all the regression models in orthonormal

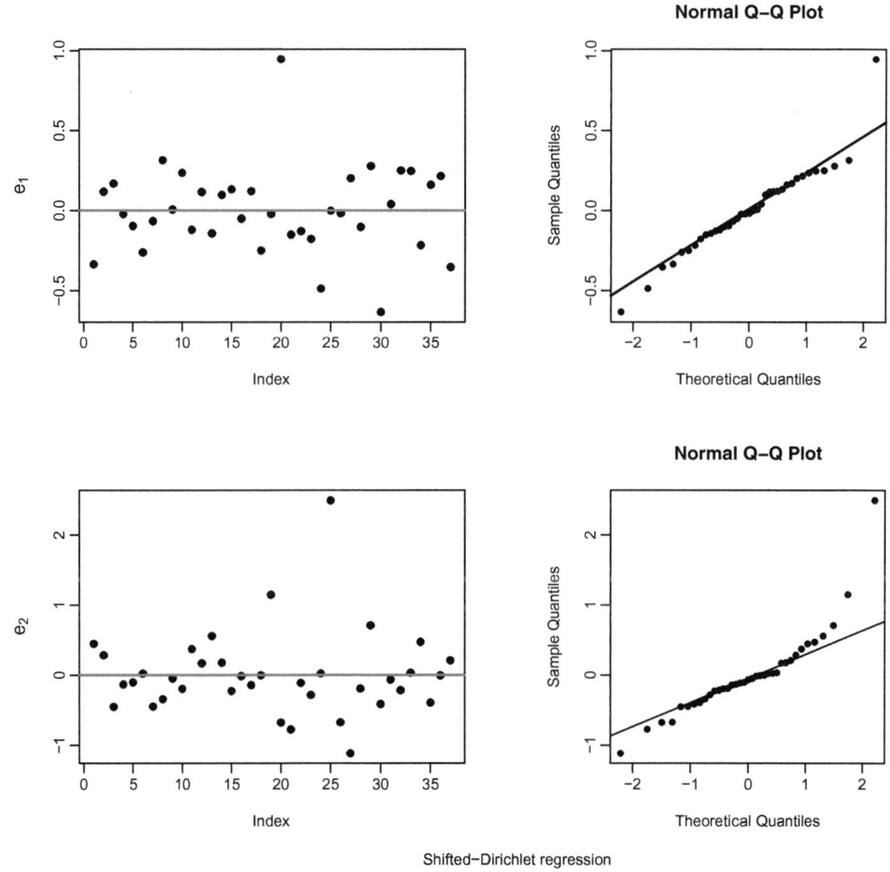

Fig. 5 In the *left column* coordinate residual plots associated to the estimated Shifted-Dirichlet regression. In the *right column* Q–Q plots for residuals of the corresponding regression models are displayed

coordinates (see Eq. (16)) which is making it possible to apply the standard battery of testing hypotheses for linear regression models, such as testing marginal normality of each coordinate residual. Coordinate residual plots and associated normal Q–Q plots for the three different regression models are displayed in Figs. 3, 4 and 5.

Except for the tails of the distribution (see Fig. 5), the assumption of normality seems to be reasonable. For the first coordinate residuals, whose points are displayed in the upper left of Fig. 5, the p-values of the Anderson–Darling test and of the Lilliefors (Kolmogorov–Smirnov) test for normality are 0.19 and 0.4789, respectively, so that the hypothesis of normal distribution cannot be rejected, while for the second coordinate residuals, lower left of Fig. 5, the two p-values of the two mentioned normality tests are 0.0009 and 0.003, respectively, due to the presence of an upper outlier (25th observation). If we omit such outlying point, normality is confirmed, i.e., the p-value of Anderson–Darling test equals 0.431 while the Lilliefors test p-value is 0.115.

5 Conclusions

Regression models with compositional response were proposed in the eighties. In this work, using the Shifted-Dirichlet distribution a new covariate model on the simplex is proposed. The Shifted-Dirichlet distribution is a generalization of the Dirichlet distribution obtained, within the Aitchison geometry, after applying a perturbation to the standard Dirichlet one. As a probability distribution, it is the same as the Scaled-Dirichlet distribution but the density is expressed with respect to the Aitchison measure on the simplex, and not with respect to the Lebesgue measure in the induced Euclidean geometry from the real space. Consequently, the Shifted-Dirichlet regression model is a generalization of the Dirichlet regression model. Even though the number of parameters to estimate increases, we see that it is a feasible and more flexible model. Using a real data set, we obtain results comparable to those obtained using the Simplicial regression.

Acknowledgments Research partially financially supported by the Italian Ministry of University and Research, FAR (Fondi di Ateneo per la Ricerca) 2012, by the Ministerio de Economía y Competividad under the projects METRICS (Ref. MTM2012-33236) and CODA-RETOS (Ref. MTM2015-65016-C2-1-R), and by the Agència de Gestió d'Ajuts Universitaris i de Recerca of the Generalitat de Catalunya under the project COSDA (Ref: 2014SGR551).

References

1. Abramowitz, M., Stegun, I.A.: Handbook of Mathematical Functions with Formulas, Graphs, and Mathematical Tables. Dover Publications, New York (1964)
2. Aitchison, J.: The Statistical Analysis of Compositional Data. Chapman & Hall Ltd. (Reprinted in 2003 with additional material by The Blackburn Press), London (1986)

3. Aitchison, J., Shen, S.M.: Logistic-normal distributions. Some properties and uses. Biometrika **67**, 261–272 (1980)
4. Billheimer, D., Guttorp, P., Fagan, W.F.: Statistical interpretation of species composition. J. Am. Stat. Assoc. **96**, 1205–1214 (2001)
5. Cameron, A.C., Windmeijer, F.A.G.: R-squared measures for count data regression models with applications to health-care utilization. J. Busin. Econ. Stat. **14**(2), 209–220 (1996)
6. Campbell, G., Mosimann, J.: Multivariate methods for proportional shape. In: ASA Proceedings of the Section on Statistical Graphics (1987)
7. Coakley, J.P., Rust, B.R.: Sedimentation in an Arctic lake. J. Sediment. Petrol. **38**, 1290–1300 (1968)
8. Egozcue, J.J., Daunis-i-Estadella, J., Pawlowsky-Glahn, V., Hron, K., Filzmoser, P.: Simplicial regression. The normal model. J. App. Prob. Stat. **6**, 87–108 (2012)
9. Egozcue, J.J., Pawlowsky-Glahn, V.: Groups of parts and their balances in compositional data analysis. Math. Geol. **37**(7), 795–828 (2005)
10. Ferrari, S., Cribari-Neto, F.: Beta regression for modelling rates and proportions. J. App. Stat. **31**, 799–815 (2004)
11. Garzanti, E., Vezzoli, G., Andó, S.: Paleogeographic and paleodrainage changes during Pleistocene glaciations (Po Plain, Northern Italy). Earth-Sci. Rev. **105**, 25–48 (2011)
12. Gueorguieva, R., Rosenheck, R., Zelterman, D.: Dirichlet component regression and its applications to psychiatric data. Comput. Stat. Data Anal. **52**, 5344–5355 (2008)
13. Hijazi, R.H.: Residuals and Diagnostics in Dirichlet regression. Tech Report of United Arab Emirates University, Department of Statistics (2006)
14. Hijazi, R.H., Jernigan, R.W.: Modelling compositional data using dirichlet regression models. J. App. Prob. Stat. **4**, 77–91 (2009)
15. Mateu-Figueras, G., Pawlowsky-Glahn, V., Egozcue, J.J.: The principle of working on coordinates. In: Compositional Data Analysis, pp. 29–42. John Wiley & Sons, Ltd (2011)
16. Mateu-Figueras, G., Pawlowsky-Glahn, V., Egozcue, J.J.: The normal distribution in some constrained sample spaces. SORT **37**(2), 231–252 (2011)
17. Melo, T.F.N., Vasconcellos, K.L.P., Lemonte, A.J.: Some restriction tests in a new class of regression models for proportions. Comput. Stat. Data Anal. **53**, 3972–3979 (2009)
18. Monti, G.S., Mateu-Figueras, G., Pawlowsky-Glahn, V.: Notes on the scaled Dirichlet distribution. In: Compositional Data Analysis, pp. 128–138. John Wiley & Sons, Ltd (2011)
19. Monti, G.S., Mateu-Figueras, G., Pawlowsky-Glahn, V., Egozcue, J.J.: The shifted-scaled Dirichlet distribution in the simplex. In: Proceedings of The 4th International Workshop on Compositional Data Analysis (2011)
20. Monti, G.S., Mateu-Figueras, G., Pawlowsky-Glahn, V. and Egozcue, J.J.: Scaled-Dirichlet covariate models for compositional data. In: Proceedings of 47th Scientific Meeting of the Italian Statistical Society (2014)
21. Pawlowsky-Glahn, V.: Statistical modelling on coordinates. In: Proceedings of Compositional Data Analysis Workshop—CoDaWork'03 (2003)
22. Pawlowsky-Glahn, V., Egozcue, J.J.: Geometric approach to statistical analysis on the simplex. Stoch Env. Res. Risk A. **15**, 384–398 (2001)
23. Pawlowsky-Glahn, V., Egozcue, J.J., Tolosana-Delgado, R.: Modelling and analysis of compositional data, p. 272. Statistics in practice. John Wiley & Sons, Chichester UK
24. R Development Core Team: R: A Language and Environment for Statistical Computing. R Foundation for Statistical Computing (2014)
25. Rayens, W.S., Srinivasan, C.: Dependence properties of generalized Liouville distributions on the simplex. J. Am. Stat. Ass. **89**, 1465–1470 (1994)

Relationship Between the Popularity of Key Words in the Google Browser and the Evolution of Worldwide Financial Indices

R. Ortells, J.J. Egozcue, M.I. Ortego and A. Garola

Abstract The purpose of this contribution is to evaluate whether there is enough statistical basis to establish a relationship between the popularity of certain terms in the Google browser and the evolution of several worldwide economic indices the subsequent week. A linear model trying to predict the evolution of 19 financial indices from all over the world with the information of how many times a selected group of 200 key words are looked up online the previous week is proposed. The linear model that is proposed takes a compositional approach due to two reasons. First, the information contained in the values of the financial indices has a compositional nature. The strongest proof supporting this idea is that in case all values for the indices on a certain week were multiplied by a factor, the information would remain unchanged. In fact, the value for a certain index is irrelevant by itself, since it is its evolution with respect to the rest of indices that indicates whether it is performing well. Therefore, this idea suggests that the numerical values of the 19 indices for a certain week can be understood as a vector of the simplex and be analyzed accordingly. Second, the explanatory variable has to be understood as a vector of the simplex as well, for a similar reason as before. For instance, let us imagine that the number of times the words are looked up online in a certain week was multiplied by a factor. Indeed, the

R. Ortells (✉)
The London School of Economics and Political Science (LSE),
London SE11 5TA, UK
e-mail: r.ortells@lse.ac.uk

J.J. Egozcue
Universitat Politècnica de Catalunya (UPC), Campus Nord. C/ Jordi Girona,
1-3, Edifici C2, Despatx 211B, 08034 Barcelona, Spain
e-mail: juan.jose.egozcue@upc.edu

M.I. Ortego
Universitat Politècnica de Catalunya (UPC), Campus Nord. C/ Jordi Girona,
1-3, Edifici C2, Despatx 307, 08034 Barcelona, Spain
e-mail: ma.isabel.ortego@upc.edu

A. Garola
Universitat Politècnica de Catalunya (UPC), Campus Nord. C/ Jordi Girona,
1-3, Edifici B1, Despatx 301, 08034 Barcelona, Spain
e-mail: alvar.garola@upc.edu

J.A. Martín-Fernández and S. Thió-Henestrosa (eds.), *Compositional Data Analysis*, Springer Proceedings in Mathematics & Statistics 187,
DOI 10.1007/978-3-319-44811-4_10

information contained in this vector would be exactly the same. Moreover, it seems intuitive as well how the absolute value for the number of searches is irrelevant by itself, since we will be interested in the relationships amongst variables. For the reasons we have just set, a compositional approach seems necessary in order to address the problem successfully, since both the explanatory and predicted variables present a compositional nature. In other words, despite not adding up to a constant, the components of the vectors of both the explanatory and predicted variables seem to be closely related in terms of giving information of a *part of a whole*, so tackling the problem through a compositional perspective seems appropriate. The analysis consists of an exploratory analysis of both response (indices) and explanatory (searches) variables and a compositional linear multiple regression between both sets of variables.

Keywords Financial markets · Google searches · Stock market indices · Compositional data · Multiple linear regression

1 Introduction

The scope of the project is to analyze whether the weekly evolution of the popularity of a group of words in the Google browser is explicative of the behavior of the financial markets on the following week. By the term *financial markets*, we understand a selection of economic indicators from all over the world, including the main stock market indices, sovereign bond yields, and commodity prices. This selection of indicators has been made assuming that the characterization of what we understand as *global markets* can be done through the most relevant financial centres of the world. To do so, we have used a data set containing the evolution of the popularity of 200 words and the performance of 19 financial indices during the period 2004–2014 (554 weeks).

The potential usefulness of the searches in the Google browser[1] has been widely studied in the past. For example, Preis et al. [25] tried to approach the correlation between the popularity of a group of selected words and the Dow Jones Industrial Average Index performance. Other economic-related lines of investigation have tried to address the correlation between the popularity of certain words in an Internet browser and other variables such as the stock market volume [7–9]. However, the information of the popularity of certain words in Google has also been used in multiple other fields apart from the economic world; for instance, Ginsberg et al. [18] focused on the potential correlation of such popularity and the spreading of epidemics. According to the state of the art, the key message is that the information on the patterns of searches in the Google browser can be explicative of the behavior of multiple indicators, as well as it can anticipate future trends. Indeed, according to the current lifestyle in most developed countries, the use of Internet seems to be

[1]The information on the searches of different words is provided by Google Trends, subsidiary of Google, Inc. From now on, we will refer to Google Trends simply as Google.

extremely useful to understand not only how society behaves at present time but also to anticipate how it will behave in the future.

With reference to what the present document intends to add to the state of the art, we have tried to address a similar problem to the one considered in Preis et al. [25] taking into account a compositional approach, both for the information of the popularity of the words as well as the economic indices. First, a justification on why such an approach is appropriate has been made. After that, a regression model has been proposed. Finally, the results have been analyzed critically, in order to evaluate whether the compositional perspective has worked as expected.

2 Explanatory and Predicted Variables

2.1 Selection and Treatment of the 200 Terms in the Google Browser

The database containing the words that have been selected does not only include economic terms. Previous research such as Preis et al. [25] exclusively considered words from the economic world; we have decided to include additional words to evaluate whether they can be explicative as well. These 200 words are presented in Table 1, and include the words that were used in Preis et al. [25] plus an additional group that has been considered by the authors. Not all the words that have been included are expected to be useful; indeed, the point of adding totally unrelated words and see that they do not have explanatory power is something we have deliberately done in order to check that the methodology works as expected. It is important to point out that we have deliberately chosen *universal* terms. For example, we have avoided using words like *subprime* or *Lehman Brothers* because even though their popularity would have certainly been explanatory of the behavior of the markets during the period 2004–2014, they would have biased the results of our project.

Regarding the information provided by Google, a few comments have to be made. First, the information on the number of searches on the browser is not the absolute number of searches of each word. For every word, the database associated to it has been *normalized* in such a way that the historical maximum appears as 100, so there is no way to know the true absolute popularity of the words. Therefore, the information of the number of searches every week is expressed as a fraction of the historical maximum. Second, the information on the number of searches has been rounded to the nearest integer value. For example, in case after the *normalization* the value associated to one week is 62.3, Google returns the rounded value 62. Provided that no further information is provided by the browser, there is uncertainty on how the raw big data has been previously treated by Google, Inc.

The database containing the information on weekly amount of searches for the 200 words has been treated in order to, first, address the compositional nature of the data, and, second, reduce the dimension according to a principal component analysis.

Table 1 List of the 200 words that have been used in the study

Debt	Society	Water	Trader	Interest rates	Savings	Shortage	Free
Color	Leverage	Rich	Rare earths	Business	Speculation	Capitalization	Year
Stocks	Loss	Risk	Tourism	Shares	Credit default swap	Utilities	University
Restaurant	Cash	Gold	Politics	Mortgage	Sovereign	Transportation	Pencil
Portfolio	Office	Success	Energy	S&P500	Treasury	Trading	Book
Inflation	Fine	Oil	Consume	Central bank	Volatility	Currency	Travel
Housing	Stock market	War	Consumption	Futures	Wage	High yield	Dog
Dow jones	Banking	Economy	Freedom	Commodities	Salary	Assets	Disco
Revenue	Crisis	Chance	Dividend	Brent	Wealth	ECB	Cinema
Economics	Happy	Short sell	World	West texas	Bank	Yen	Bus
Credit	Car	Lifestyle	Conflict	Equity	Exports	Yuan	Animal
Markets	NASDAQ	Greed	Kitchen	Tax	Imports	Securities	Safari
Return	Gains	Food	Forex	Deflation	Income	Warrants	Name
Unemployment	Finance	Financial markets	Home	GDP	Capital	ETF	Think
Money	Sell	Movie	Crash	IPO	Hedge fund	CFD	Boy
Religion	Investment	NYSE	Transaction	Rating	Loan	Insolvency	Old
Cancer	Fed	Ore	Garden	Bund	Eurozone	Disaster	Word
Growth	House	Opportunity	Fond	Bankruptcy	Euro	Fund	Number
Investment	Metals	Health	Train	Bullish	Dollar	Happiness	Press
Hedge	Travel	Short selling	Labor	Bearish	BRIC	Sex	Mother
Marriage	Returns	Earnings	Fun	Productivity	Terrorism	Church	Sun
Bonds	Gains	Arts	Environment	Deficit	Attack	Shopping	Eye
Derivatives	Default	Culture	Ring	Devaluation	Collapse	Computer	Life
Headlines	Present	Bubble	Recession	Federal reserve	G20	Science	Time
Profit	Holiday	Buy	Yield	Liquidity	Danger	Film	Person

First of all, the zeros on the database have been identified and addressed in the way that was proposed by Martín-Fernández et al. [21]. The decision of approaching the presence of zeros in this way comes from the strong belief that such zeros are *rounding zeros* (very small amount of searches compared to the historical maximum that appear as zeros even though they are not). After dealing with the zeros on the database, a compositional approach is considered. There are many explanations supporting the compositional nature of the database. First, in case the number of searches were all multiplied by a factor k, the information would remain unchanged. Second, the information for a certain word is irrelevant by itself since it is only explicative when compared to the rest. For instance, in case the word *bankruptcy* increases in popularity for a certain week and the other words present a similar increase, it seems that not much has happened. However, in case the searches increase for such word and remain quite the same for the rest, this may be relevant. For this reason, it seems appropriate to understand the number of searches for a certain word as a *part of a whole* and apply a compositional treatment accordingly. However, we should be aware that the sum of the searches for a certain week is not constant, and is unknown. Indeed, the total number of times that the 200 words are looked up is different every week, but this is irrelevant because we are not interested in working with the raw data. It is important to highlight that this detail makes the present study slightly different from the traditional compositional analysis, since even though it seems intuitive to understand the information as parts of a whole the *actual whole* is not only variable for each week but also unknown.

Once the compositional treatment has been justified, we have addressed the fact that the information on the number of searches has been normalized in such a way that the historical maximum has become 100. For a certain week i, we have a 200 component vector containing the information of the popularity of each word:

$$X_i = [x_1, ..., x_{200}].$$

We have already proved how this vector belongs to the simplex space [15, 23] due to its compositional nature. However, we can see how each component of the vector represents the absolute number of searches multiplied by an unknown *normalizing factor* such that the historical maximum becomes 100. This operation is equivalent to a perturbation in the simplex for each one of the 200 components, so the relative information contained in the vector remains undamaged (for further information on how perturbation does not alter the information of the vector see Aitchison [1, 2] and Pawlowsky-Glahn et al. [23]). It is important to point out that the value for this normalizing factor is irrelevant, since we are not interested in obtaining the absolute number of searches.

The compositional treatment has consisted in computing the CLR coordinates (centered logratio coordinates) of the dataset, according to Pawlowsky-Glahn et al. [23]. After that, we have applied a singular value decomposition and obtained the principal components of the space. The ultimate scope of performing such operation is to reduce the dimension of the vector so we can decrease the number of explanatory variables in the regression. Regarding the regression, we will only consider the first

Table 2 Cumulated explained variance by the first 20 principal components of the data set

Component	Cumulated variance (%)	Component	Cumulated variance (%)
Comp. 1	53.3	Comp. 11	89.5
Comp. 2	65.2	Comp. 12	90.4
Comp. 3	72.3	Comp. 13	91.3
Comp. 4	76.7	Comp. 14	92.0
Comp. 5	79.6	Comp. 15	92.6
Comp. 6	82.2	Comp. 16	93.1
Comp. 7	84.2	Comp. 17	93.5
Comp. 8	85.8	Comp. 18	93.9
Comp. 9	87.2	Comp. 19	94.3
Comp. 10	88.4	Comp. 20	94.6

Source Results of the principal component analysis made on the data with R software

15 components of the vectors resulting from the principal component analysis. The percentage of the cumulated explained variance up to the 20th component is detailed in Table 2.

To sum up, the database containing the information on the searches of the words has been treated as compositional (through a CLR transformation[2]), and has been simplified according to a singular value decomposition criterion. The final result has been a new database, where each week is characterized through a vector of 15 components (since we have limited the number of explanatory variables to the 15 first components of the vectors), which will be our explanatory variables in the regression model we will propose. All computations have been made with R software.

2.2 *Selection and Treatment of the Indices*

The evolution of the financial markets has been modeled through studying 19 key indices, which are representative of stock market indices and sovereign bond yields from all over the world, as well as the most important commodity markets. The information on the values of the indices has been obtained from Yahoo Finance[3] and Investing.com.[4] In fact, in order to make a thorough analysis of the worldwide financial markets we could have taken into consideration every stock market, fixed

[2]It is arguable whether we could have used an ILR transformation, as we have done for the predicted variable. Provided that a principal component analysis will be applied, it is unnecessary to define a binary partition and an ILR transformation. Given that in the end we will work with the orthogonal base that we get from the principal component analysis, we can work directly with a standard CLR procedure.

[3]Yahoo Finance, owned by Yahoo!, Inc. Information consulted in 2013.

[4]Investing.com, owned by Fusion Media Limited. Information consulted in 2014.

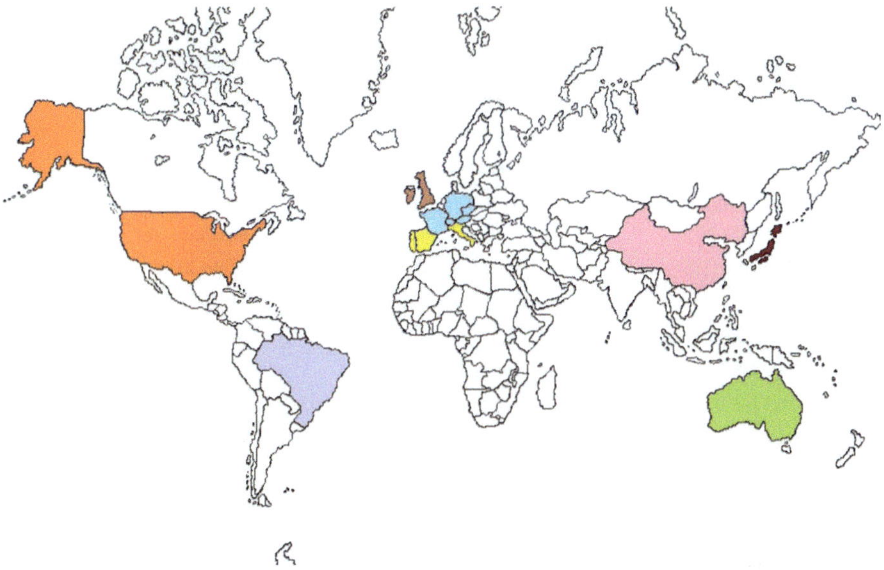

Fig. 1 World map with the main areas that have been studied through the selected market indices

income index and commodity that is currently traded in the planet. By doing this, we would certainly be able to *see* how the worldwide resources move along what we understand as *global markets*. However, the difficulty of doing so is that this methodology would be prohibitively expensive in terms of computational cost; therefore, a simplification has had to be made.

The approach we have taken consists in considering the major financial centres of the world. This simplification is equivalent to assume that the evolution of the worldwide financial markets can be modeled through the stock exchanges with higher trading volume on the planet. Even though such assumption may not be strictly true, it seems a good simplification of the reality, since any major change in the evolution of any market in the world will certainly affect these major financial *hubs*, so it will be perceived somehow in our model. In Fig. 1 we present the countries that are directly studied through the selected financial indices. A list containing the 19 indices is presented in Table 3.

The information of the indices has been converted into one single currency, US Dollar, in order to have a trustworthy representation of the price of the indices. Such transformation has been done through the spot currency exchange (value for the currency exchange in the foreign exchange market at the end of the week). In case we did not do it, the information provided by the indices would be partial, since the actual *value* of the index would depend on the exchange currency rate (which varies every day). Once this dollarization has been carried out, we can discuss the compositional nature of the indices.

Table 3 List of the 19 indices that have been considered for the study

Name of the index	Location of the market	Currency of issue
Dow Jones Industrial Average (DJIA)	New York, USA	US Dollar (USD)
Eurostoxx50 (EUSTX)	Frankfurt, Germany	Euro (EUR)
Milano Italia Borsa (MIB)	Milano, Italy	Euro (EUR)
Footsei100 (FTSE)	London, UK	Great Britain Pound (GBP)
Nikkei225 (NIKKEI)	Tokyo, Japan	Japanese Yen (JPY)
Hang Seng Index (HSI)	Hong Kong, Hong Kong	Hong Kong Dollar (HKD)
Bovespa	Sao Paulo, Brazil	Brazilian Real (BRL)
Australian Stock Exchange (ASX)	Sydney, Australia	Australian Dollar (AUD)
10 year bond USA	USA	US Dollar (USD)
10 year bond Japan	Japan	Japanese Yen (JPY)
10 year bond Germany (BUND)	Germany	Euro (EUR)
10 year bond Spain	Spain	Euro (EUR)
10 year bond Hong Kong*	Hong Kong	US Dollar (USD)
10 year bond UK	UK	Great Britain Pound (GBP)
10 year bond Australia	Australia	Australian Dollar (AUD)
Gold (futures due earliest date)	New York (NYMEX)	US Dollar (USD)
Brent (futures due earliest date)	London (ICE)	US Dollar (USD)
Cocoa (futures due earliest date)	London (ICE)	US Dollar (USD)
Corn (futures due earliest date)	London (ICE)	US Dollar (USD)

*In fact, the index we have used is not explicitly the 10 year Hong Kong bond, we have used an index containing a group of Asian bonds issued in US Dollars. However, the evolution of such index can be understood as the evolution of the fixed income market in that region, though not being strictly Chinese

The key argument supporting the fact that the information of the indices is compositional is that in case these indices were multiplied by a factor k, the information would remain unchanged. In other words, the actual numeric value of an index is irrelevant by itself, the important information is how it performs with respect to the rest of indices. However, even though the set of values for the 19 indices can be understood as a vector of the simplex, the sum is not constant, so our compositional approach will slightly differ from the classical compositional analysis, thoroughly explained in Aitchison [2] and Egozcue and Pawlowsky-Glahn [16]; provided that the actual sum of the components is unknown and variable we will not be able to define the *closure* operation. Anyway, the fact of not being able to perform the closure operation, which requires to know the actual value for the sum, will not be necessary, since we will be exclusively interested in the evolution of the indices with respect to the rest.

It is especially relevant to highlight that since the sum of the indices is variable and unknown we will not be able to predict *actual numeric values* for the indices. From a traditional point of view where the most relevant information of the model is the predicted value for every index, this may seem disappointing; it is certainly not,

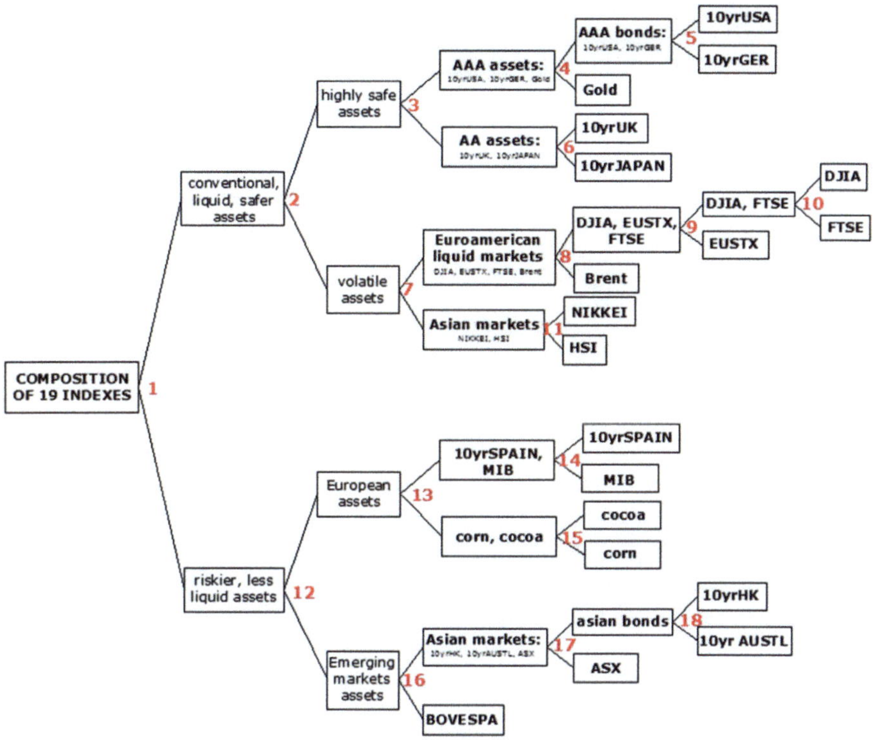

Fig. 2 Scheme of how the 19 indices have been separated in order to create the binary partition needed to compute the ILR coordinates. The divisions have been numbered in red so the corresponding ILR coordinate is easy to identify

because in case such algorithm was used for trading purposes, with the information on the relative expected behavior of the indices we would be able to get a position in the markets.

The compositional approach has consisted in defining a sequential binary partition and computing the corresponding ILR coordinates (isometric logratio coordinates), as it was developed in Egozcue et al. [17] and Egozcue and Pawlowsky-Glahn [14]. The definition of such partition is not random; in fact, according to the information that we had a priori, we have created a partition with several ratios that are of interest. This partition is presented schematically in Fig. 2. Before moving onto the definition of the coordinates, a few remarks have to be done regarding why we have considered such divisions. First, we have distinguished between two types of indices: in one group we have the traditional, highly liquid indices, which are present nearly in every portfolio of any investment firm of the world. These indices include the high quality sovereign bonds (bonds issued by the US, Germany, the UK, or Japan) as well as the most liquid stock market indices (New York, London, Frankfurt, Japan, and Hong Kong). We have also included in this group the Brent and Gold futures because they represent highly liquid assets that are very common as well in any investment portfolio. In the other group, we have included the rest of indices, including Southern Europe assets as well as Emerging Markets indicators, such as the Sao Paulo Stock Exchange. It is especially relevant to point out that even though alternative divisions based on, for example, the location of the indices, would have been correct as well, they would not have represented this clear binomial between high-quality assets versus riskier assets, which is the one assumed to be closely related to the Google searching pattern. The underlying hypothesis under this procedure is that investors are risk averse: in other words, risky assets will decrease in value when uncertainty arises because their expected returns will not compensate any more the associated risk that must be taken when investing in them (for further information on risk aversion see Arrow [5] and Pratt [24]).

Once this first division has been made, we have proceeded similarly for each group. For instance, we have divided the high-quality assets group into two subgroups, safer assets versus stock market indices. After that, we have made subsequent divisions according to the same criterion. In the same way, we have divided the indices contained in the riskier assets group. Overall, the key message is that we have tried to separate the indices in such a way that makes sense in terms of the decision that any rational investor would have to make. For example, in case the perspectives for the future worsen, any rational investor would decide to invest in highly liquid, conventional assets, and liquidate any position in the riskier, more volatile indices. Therefore, the partition that we have to create has to represent such decision in order to be meaningful.

To sum up, the treatment of the financial indices through an ILR coordinate transformation has been performed because the nature of such data expresses relative information. In order to do so, a strategic sequential binary partition has been defined and the standard procedure to compute the ILR coordinates that was proposed by Egozcue et al. [17] has been applied. The information containing the sequential binary partition is presented in Table 4. These computations have been made through Codapack and R software.

After having defined the corresponding treatments on both data sets, an exploratory analysis has been performed. We present the compositional biplots that have been built in Figs. 3 and 4 (more information on compositional biplots in Aitchison and Greenacre [3]). It is especially relevant to point out how both biplots appear to follow a cyclical pattern: it seems that both the explanatory and response variables evolve in a way where there is a somewhat cyclical pattern. However, there is something that makes the situations over time different from the past. Indeed, the fact of observing a cyclical pattern in economic data is something we might have expected, yet the specific shape of the biplots remains unexplained.

Table 4 Sequential binary partition that has been used to create the ILR coordinates for the indices

	DJIA	NIK	MIB	EUSTX	ASX	BOV	HSI	FTSE	10US	10JAP	10SPA	10GER	10AUS	10HK	10UK	Brent	Gold	Corn	Cocoa
ILR1	1	1	−1	1	−1	−1	1	1	1	1	−1	1	−1	−1	1	1	1	−1	−1
ILR2	−1	−1	0	−1	0	0	−1	−1	1	1	0	1	0	0	1	−1	1	0	0
ILR3	0	0	0	0	0	0	0	0	1	−1	0	1	0	0	−1	0	1	0	0
ILR4	0	0	0	0	0	0	0	0	1	0	0	1	0	0	0	0	−1	0	0
ILR5	0	0	0	0	0	0	0	0	1	0	0	−1	0	0	0	0	0	0	0
ILR6	0	0	0	0	0	0	0	0	0	−1	0	0	0	0	1	0	0	0	0
ILR7	1	−1	0	1	0	0	−1	1	0	0	0	0	0	0	0	1	0	0	0
ILR8	1	0	0	1	0	0	0	1	0	0	0	0	0	0	0	−1	0	0	0
ILR9	1	0	0	−1	0	0	0	1	0	0	0	0	0	0	0	0	0	0	0
ILR10	1	0	0	0	0	0	0	−1	0	0	0	0	0	0	0	0	0	0	0
ILR11	0	1	0	0	0	0	−1	0	0	0	0	0	0	0	0	0	0	0	0
ILR12	0	0	1	0	−1	−1	0	0	0	0	1	0	−1	−1	0	0	0	1	1
ILR13	0	0	1	0	0	0	0	0	0	0	1	0	0	0	0	0	0	−1	−1
ILR14	0	0	1	0	0	0	0	0	0	0	−1	0	0	0	0	0	0	0	0
ILR15	0	0	0	0	0	0	0	0	0	0	0	0	0	0	0	0	0	−1	1
ILR16	0	0	0	0	1	−1	0	0	0	0	0	0	1	1	0	0	0	0	0
ILR17	0	0	0	0	−1	0	0	0	0	0	0	0	1	1	0	0	0	0	0
ILR18	0	0	0	0	0	0	0	0	0	0	0	0	−1	1	0	0	0	0	0

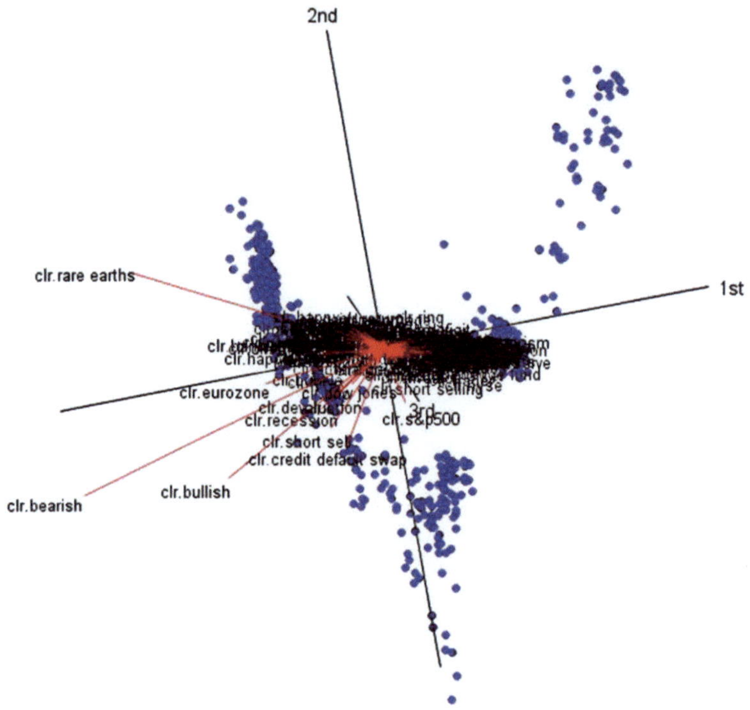

Fig. 3 Compositional biplots of the data containing the popularity of the 200 words, built through Codapack software

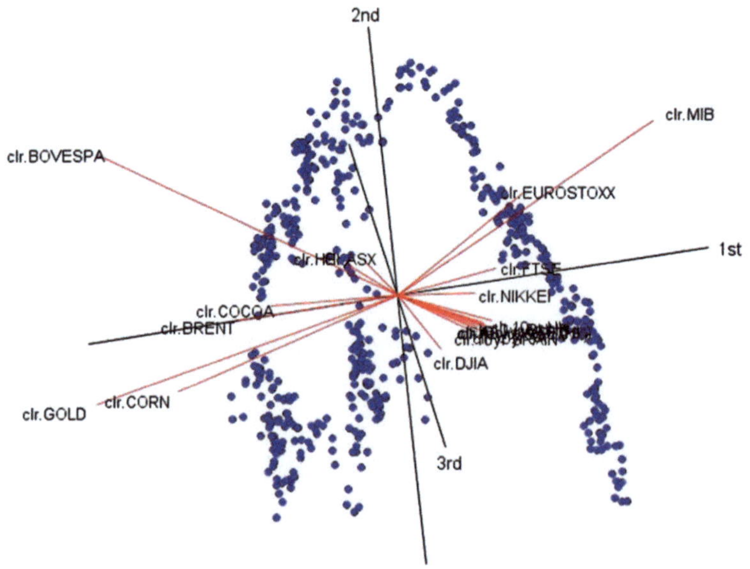

Fig. 4 Compositional biplots of the data containing the 19 financial indices, built through Codapack software

3 Regression Model

The model that has been built consists in a multiple linear OLS regression, where both the explanatory and predicted variable have had a previous compositional treatment. This model has been built with R software. In total, we have predicted 18 ILR coordinates with a linear combination of 15 variables (which are the first 15 principal components of the vector containing the CLR transformation of the searches). The expression of the model is presented in Eq. 1.

$$ILR_j = \beta_{0j} + \sum_{i=1}^{i=15} \beta_{ij} x_i, \quad \forall j \in [1, 18]. \tag{1}$$

The normality on the residuals has been analyzed on the 18 regressions, through carrying out the Kolmogorov-Smirnov test [22], the Anderson-Darling test [4] and the Shapiro-Wilk test [26]. Since we are dealing with a large database, the KS test is not appropriate because it rejects the normality hypothesis easily (for further information on the suitability of the KS test for large databases see Babu and Rao [6]). The results of the corresponding p-values have been presented in Table 5. We have rejected the hypothesis of normality on the residuals unless both the AD and SW test have provided p-values above 0.05 and 0.10, respectively (limits imposed according to Anderson and Darling [4] and Royston [26]). We have concluded that normality on the residuals can be rejected for all the regressions except ILR7, ILR8, and ILR11.

It is relevant to point out a few details with regard to the results of the regression. First, recall that no collinearity may be expected, since the variables come from a principal component decomposition so they are orthogonal by definition. Second, no leverage problems have been observed due to the data, by evaluating all the points of the model with the Cook distance, according to Cook [10] and Kim and Storer [19]. Finally, potential autocorrelation patterns on the residuals have been evaluated through the Durbin–Watson test [11, 12], and the conclusion has been that no first order autocorrelation may be expected from the present model. However, the fact that we are dealing with econometric time series suggests that more complex autocorrelation patterns might be present.

The goodness of fit of the model has been evaluated through the R^2 coefficient. The values for the R^2 are presented in Table 6. Provided that we are fitting 554 points with a 15 variable model the results seem highly positive. It would be arguable whether 15 variables are too many, and whether a simplified model would be more appropriate. Indeed, since we are performing 18 regressions and each one requires fitting 16 coefficients, our model, which was aimed to be simple, demands nothing less than 288 coefficients. However, since the database we count on is considerably large the present solution seems adequate.

A relevant conclusion of the results is that the F test we have performed in the 18 regressions has proven the linear model we have defined makes sense; provided that the p-value for the F test in all 18 cases is $2.2 \cdot 10^{-16}$ we can infer that the modeling of the evolution of the financial indices through an isometric log-ratio transformation

Table 5 Results of the tests checking the normality of the residuals

Coordinate	p-value KS	p-value AD	p-value SW	Normality
ILR1	2.20E−16	0.000	0.000	NO
ILR2	2.20E−16	0.009	0.003	NO
ILR3	2.20E−16	0.052	0.057	NO
ILR4	2.20E−16	0.069	0.058	NO
ILR5	2.20E−16	0.000	0.003	NO
ILR6	2.20E−16	0.000	0.000	NO
ILR7	2.20E−16	0.528	0.203	YES
ILR8	2.20E−16	0.338	0.354	YES
ILR9	2.20E−16	0.003	0.000	NO
ILR10	2.20E−16	0.002	0.000	NO
ILR11	2.20E−16	0.141	0.108	YES
ILR12	2.20E−16	0.032	0.121	NO
ILR13	2.20E−16	0.076	0.023	NO
ILR14	2.20E−16	0.000	0.003	NO
ILR15	2.20E−16	0.122	0.073	NO
ILR16	2.20E−16	0.000	0.000	NO
ILR17	2.20E−16	0.000	0.000	NO
ILR18	2.20E−16	0.098	0.002	NO

The criterion to reject the hypothesis on the normality of the residuals is that either the Anderson–Darling (AD) or the Shapiro–Wilk (SW) test provides a p-value below the significance level of 0.05 and 0.10, respectively. The Kolmogorov–Smirnov (KS) test has proved to be too strict when dealing with large databases, so its result has not been considered. Hypothesis of normal distribution on the residuals rejected for all regressions except ILR7, ILR8, and ILR11
Source Computations made through R software

Table 6 Goodness of fit indicators for the regression on the 18 coordinates

Coordinate	R^2	Coordinate	R^2
ILR1	0.801	ILR10	0.889
ILR2	0.791	ILR11	0.609
ILR3	0.950	ILR12	0.933
ILR4	0.949	ILR13	0.838
ILR5	0.724	ILR14	0.385
ILR6	0.844	ILR15	0.872
ILR7	0.507	ILR16	0.773
ILR8	0.818	ILR17	0.740
ILR9	0.845	ILR18	0.660

Source Results obtained through computations with R software

(ILR) is undoubtedly meaningful. Several ideas can be confirmed from such strong results. First, modeling the indices through a binary partition approach is appropriate. Even though we had thoroughly justified why the relevant information on the indices is on the *relative performance* with respect to the rest rather than on their absolute evolution, we confirm that this intuitive idea is correct. Second, we can also confirm that the compositional approach we have taken for the information on the popularity of the words is appropriate as well. Indeed, even though the compositional treatment looked adequate according to the nature of the data, we can confirm that by proceeding this way the model is meaningful. Last, we have also confirmed that a linear model is a correct approximation. In fact, even though both compositional treatments on the variables seemed reasonable, we did not have any information on whether a linear model would be enough, which has been the case.

4 Discussion of the Results

Even though the 18 regressions we have proposed have turned out to be meaningful, the fit coefficients R^2 differ, with values that oscillate between 0.385 and 0.95. It is especially relevant to address such differences and whether this is a situation we could have expected. There are three regressions with a fit coefficient above 0.93, and provided that we are adjusting 554 points this is an extremely powerful result. The balances associated to these regressions are the following comparisons: triple A assets versus double A assets, triple A bonds versus Gold and European risky assets versus Emerging Markets assets.

If we take a closer look we can see that it is extremely intuitive that these particular ratios are fitted better than the rest with the data from the Google searches, since they represent relationships between assets that behave *systematically* in the same way regardless of the time period we consider. Let us consider the example, where there is a sharp increase in uncertainty and turmoil arises in the markets. In that case, the AAA assets will recurrently perform better than AA assets, so the ratio between them will increase (unless AAA assets stop being more secure than AA). However, in case we think of another ratio different from the previous ones, the behavior is not that predictable; for instance, in case turmoil arises it is not clear whether the Nikkei (Tokyo Stock Market) will perform better than the Hang Seng (Hong Kong Stock Market), so it seems reasonable that the R^2 drops to 0.609. Therefore, from the results we obtain from the R^2 we can infer that there are several combinations of assets that recurrently in a behave predictable way when certain situations happen, whereas there are other combinations of assets where the behavior is not easily predictable.

On the other extreme of the goodness of fit, we have the ratio between the Milano Stock Exchange and the 10 year Spanish bond, where the R^2 is 0.385. This seems to prove that the popularity of the terms in Google cannot explain the relationship between these two types of assets, or in case it does the goodness of fit is not accurate. This is a result that might seem shocking in the beginning, but it is not. It is widely assumed that there is an inverse correlation between the stock markets and the sovereign bond yields: once uncertainty arises, funds tend to move to safer assets (bonds), and avoid risky assets (stocks) until confidence is restored. Therefore, it would seem sensible to expect a good goodness of fit in the regression, since bonds should *systematically* behave better than stocks at rough periods and the other way

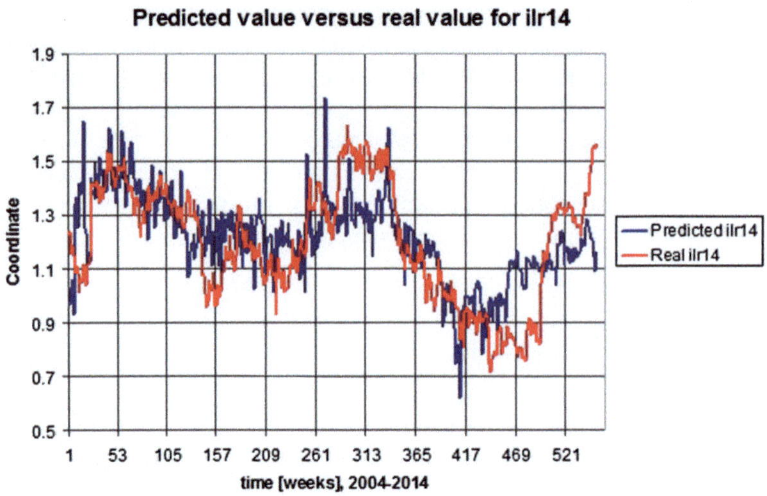

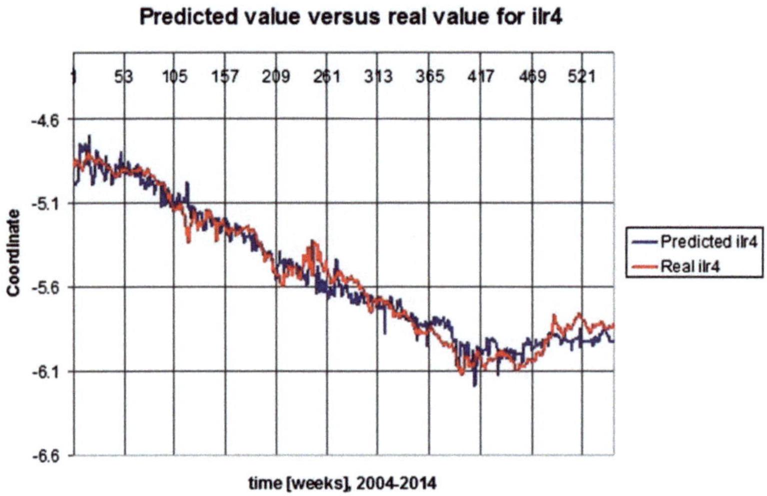

Fig. 5 Evolution of the ILR14 and ILR4 coordinates, corresponding to the ratios involving the Milano Stock Exchange versus the 10 year Spanish bond yield and the AAA assets performance versus AA assets in the time period 2004–2014. Each year has been marked with a vertical line (each year is equivalent to 52 weeks)

around in case the optimism is restored. However, that is not what happened in Southern Europe in the time period we have studied. In Fig. 5 we have presented the evolution of ILR14 (Milano Stock Exchange versus 10 year Spanish bond) and the evolution of ILR4 (AAA assets versus AA assets). If we look closely at the evolution of Milano stocks versus Spanish bonds we will see that there are localized discrepancies between the fitted values and the real values: in other words, it is not that the model consistently fits badly the data, the fact is that there are certain time periods where the prediction is clearly different from the real situation. If we look carefully at these discrepancies, we can easily understand what has happened. The first discrepancy arises in 2010, which corresponds to the Eurozone debt crisis: in that situation, the model gets information of uncertainty, and predicts that bonds should perform better than stocks, but this is not the case, since in this period Southern Europe bonds were assumed to be extremely risky as well, so all the funds were escaping *both* from the stock market and the sovereign bond market.

The second result we can comment on is that despite having different values for the R^2 on the 18 regressions, the information from the Google browser users is clearly explicative of the behavior of the relationships amongst different market indicators the following week, even though the knowledge of such users on these market indicators might be (very) limited. In other words, even though the Google users may not be individually able to infer an accurate prediction on how, for example, AAA assets will perform with respect to Gold the subsequent week, they collectively provide an extremely good approximation of such behavior. This phenomenon was first introduced in Surowiecki [27] and referred as *Wisdom of Crowds* (for further information on this concept see Surowiecki [28] and Koohang et al. [20]). The present project has not been aimed to address this phenomenon, yet it has proven that the information of the popularity of certain words that are looked up by people who not necessarily have thorough financial background is highly correlated with the performance of several economic indices the subsequent week. It has not been the purpose of the present study to assess whether a cause-effect relationship can be established. Rather, we can exclusively confirm that there is strong correlation.

Regarding the accuracy of the results, it has been found that there are several coordinates that are extremely well described through the Google search patterns, while there are others that appear to be quite independent. As misleading as this might have seemed in the first place, it has been proved how this is something that in fact might have been anticipated. On the one hand, there are certain coordinates (we should recall that coordinates are nothing more than a logratio between indices) that reproduce extremely well the flow of capital that appears when perception of risk is modified. These are the coordinates that present highly accurate fit coefficients, since they recurrently behave in the same way when similar circumstances arise. The clearest example of that is the coordinate that stands for the ratio of AAA assets versus AA assets ($R^2 = 0.95$); as long as AAA assets are more secure than AA ones, investors will always prefer them in times of financial turmoil and dislike them as soon as the perception of risk disappears. On the other hand, there are other coordinates where the ratio does not reflect as clearly the pattern we have just described, since it is not straightforward to infer whether the indices on the numerator will outperform the ones on the denominator in times of volatility. The example that has been used to exemplify this fact is the coordinate ILR11, which is the ratio between the Tokyo Stock Exchange market (Nikkei225) and the Hong Kong Stock Exchange market (Hang Seng Index). Indeed, it seems unclear to determine whether one or the other will perform better in times of financial instability. In fact, it could be the case that depending on the nature of the crisis it is one or the other which behaves better. For this reason, as long as this coordinate does not reflect the flow of capital that moves according to the risk that investors perceive, it is highly reasonable to see that the correlation with the popularity of the words in Google is weaker than in other cases.

Perhaps the most interesting conclusion has been to acknowledge how accurate the information of Google searches can be with respect to future evolution of certain relationships between financial indices. Indeed, it is remarkably interesting to see how Google users, who might have very limited knowledge on the evolution of worldwide financial indices, can collectively build up a highly accurate predictor. We have to say, however, that, this tool has been proven extremely powerful *on average*: this means that Google search patterns have recurrently predicted well the primary trend of the coordinates, even though they do not offer protection against variance. Indeed, the fact of considering weekly periods makes predictions inaccurate in terms of value, but consistently good in average terms. The ultimate implication of this conclusion is that this would probably a poor tool in case it was used for trading purposes, even though there has been clear evidence of association between variables.

It is especially relevant to focus on the nature of the wisdom of crowds, since their potential implications are highly relevant. If we stop to think about all information that is currently available with respect to economic activity all over the world, we will realize that it is virtually impossible to gather it all, nor updating it at the pace it keeps appearing. For that reason, we shall conclude that it is not humanly possible to have complete information on the evolution of the financial markets in all its global dimension, so any opinion or knowledge an individual investor might have is undeniably limited and biased by his own personal situation. The fact that a *collective network*, who has incomplete information on the situation of the markets, is *smarter* than potentially experienced individual investors seems to be, at least, not surprising. It is true that a lot can be said on the fact that individual actors might be able to protect themselves from variance; however, the fact that the model is dealing with weekly information should be kept in mind. Overall, the results of the present project have merely brought up that the Google community, which is formed by extremely diverse people (in terms of culture, background, education, and income level) has been proven capable of predicting on average what the immediate future is going to look like in the financial markets.

The model that has been proposed so far has tried to establish whether there is a correlation between the two data sets of study. To do so, and taking into consideration the amount of points and the dimension of the variables, a dimension reduction has been considered in the explanatory variable through a principal component analysis. However, such reduction has been made in the way that the minimum variance in the data is lost. In other words, the selection of the components for the regression has been chosen in order to lose the least amount of variance in the data, not to maximize the goodness of fit of the predicted variable. Therefore, even though the information we have eliminated through the reduction might not be relevant with respect to the variance of the data set, it might be counterproductive with respect to the results of the regression. There has been some research on this aspect, for example Efron et al. [13] proposed an alternative way to build regression models, called least angle regression. Even though we have not explored thoroughly these options, they are unable to tackle a dimension reduction, so even though they might seem conceptually desirable techniques, they fail to address the excessively large dimension of the explanatory variable.

5 Conclusions

The present project has approached the potential correlation between the popularity of several key words in the Google browser and the evolution of a selection of financial indices which involve equities, sovereign bond yields and commodities during the following week. However, a prior compositional treatment has been proposed on both explanatory and predicted variables, since both data sets presented a compositional nature (relative information, scale invariance, variables informative of a *part of a whole*).

Regarding the results, there appears to be strong correlation between both data sets, even though there is no evidence on whether this association can be explained through a cause-effect relationship. As long as this association is acknowledged, we might infer that the Google community search patterns are highly related to how financial markets are going to perform in the subsequent week. This phenomenon has been associated to the concept *wisdom of crowds*, which proves that even though the information that any individual user of Google might have regarding the economic situation might be limited (and even biased), the whole Google community is an extremely powerful indicator of the current and immediate future performance of the most important financial indices.

With reference to the compositional treatment, we have explored an example in which the sum of the components of the vectors is not constant (and it is unknown). It would be relevant to highlight that in this particular case the closure operation is unavailable, and it is not possible to bring back the results to the simplex. Still, this situation does not generate any methodological problem in terms of how the compositional approach is performed. Regarding possible future research, the stability of the relationships through time should be assessed. For example, by adding future data and evaluating whether the model is modified.

Acknowledgments The present research work has been developed during the time that Robert Ortells has benefited from a grant within the program "Becas de Colaboración" of the Spanish Ministerio de Educación, for the development of the project "Correlació entre l'evolució dels mercats financers globals i la popularitat de diferents paraules al buscador Google" during 2014–2015. This project has been developed in the Applied Mathematics III Department of Universitat Politècnica de Catalunya (UPC). This research has been also partially funded by the Spanish Ministerio de Economía y Competitividad under project 'Metrics' Ref.MTM2012-33236, and by the Agència de Gestió d'Ajuts Universitaris i de Recerca (AGAUR) of the Generalitat de Catalunya under the project "Compositional and Spatial Analysis" (COSDA) (Ref: 2014SGR551;2014-2016).

References

1. Aitchison, J.: The statistical analysis of compositional data (with discussion). J. R. Stat. Soc. Ser. B (Stat. Methodol.) **44**(2), 139–177 (1982)
2. Aitchison, J.: The Statistical Analysis of Compositional Data (Reprinted in 2003 by The Blackburn Press), p. 416. Chapman & Hall Ltd., London (UK) (1986)
3. Aitchison, J., Greenacre, M.: Biplots of compositional data. J. R. Stat. Soc. Ser. C (Appl. Stat.) **51**(4), 375–392 (2002)
4. Anderson, T.W., Darling, D.A.: Asymptotic theory of certain goodness-of-fit criteria based on stochastic processes. Ann. Math. Stat. **23**, 193–212 (1952)
5. Arrow, K.J.: Aspects of the theory of risk bearing. The theory of risk aversion. Helsinki: Yrjo Jahnssonin Saatio, Reprinted in: Essays in the theory of risk bearing, Markham Publ. Co., Chicago, 1971 (1965)
6. Babu, G.J., Rao, C.R.: Goodness-of-fit tests when parameters are estimated. Technometrics (Am. Stat. Assoc.) **66**, 63–74 (2004)
7. Bordino, I., Battiston, S., Caldarelli, G., et al.: Web search queries can predict stock market volumes. PloS one **7**(7):e40, 014 (2012)
8. Challet, D., Marsili, M., Zhang, Y.C., et al.: Minority Games: Interacting Agents in Financial Markets. OUP Catalogue (2013)
9. Choi, H., Varian, H.: Predicting the present with Google trends. Econ. Rec. **88**, 2–9 (2012)
10. Cook, R.D.: Detection of influential observations in linear regression. Technometrics (Am. Stat. Assoc.) **19**, 15–18 (1977)
11. Durbin, J., Watson, G.S.: Testing for serial correlation in least squares regression. I. Biometrika **37**, 409–428 (1950)
12. Durbin, J., Watson, G.S.: Testing for serial correlation in least squares regression. II. Biometrika **38**, 159–179 (1951)
13. Efron, B., Hastie, T., Johnstone, I., et al.: Least angle regression. Ann. Stat. **32**(2), 407–499 (2004)
14. Egozcue, J.J., Pawlowsky-Glahn, V.: Groups of parts and their balances in compositional data analysis. Math. Geol. **37**(7), 799–832 (2005)
15. Egozcue, J.J., Pawlowsky-Glahn, V.: Compositional data and their analysis: an introduction. Geol. Soc., Lond., Spec. Publ. **264**, 1–10 (2006)
16. Egozcue, J.J., Pawlowsky-Glahn, V.: Simplicial geometry for compositional data. Geol. Soc., Lond., Spec. Publ. **264**, 145–159 (2006)
17. Egozcue, J.J., Pawlowsky-Glahn, V., Mateu-Figueras, G., et al.: Isometric logratio transformations for compositional data analysis. Math. Geol. **35**(3), 279–300 (2003)
18. Ginsberg, J., et al.: Detecting influenza epidemics using search engine query data. Nature **457**, 1012–1014 (2009)
19. Kim, C., Storer, B.E.: Reference values for Cook's distance. Commun. Stat. Simul. Comput. **25**, 691–708 (1996)
20. Koohang, A., Harman, K., Britz, J.: Knowledge Management: Theoretical Foundation (Chapter 6: Network Analysis and Crowds of People as Sources of New Organisational Knowledge). Informing Science Press, Santa Rosa, CA, US (2008)
21. Martín-Fernández, J.A., Hron, K., Templ, M., et al.: Model-based replacement of rounded zeros in compositional data: classical and robust approach. Comput. Stat. Data Anal. **56**, 2688–2704 (2012)
22. Massey, F.J.: The Kolmogorov-Smirnov test for goodness of fit. J. Am. Stat. Assoc. **46**(253), 68–78 (1951)
23. Pawlowsky-Glahn, V., Egozcue, J.J., Tolosana-Delgado, R.: Modeling and analysis of compositional data. Wiley (2015)
24. Pratt, J.W.: Risk aversion in the small and in the large. Econometrica **32**(1–2), 122–136 (1964)
25. Preis, T., Moat, H.S., Stanley, H.E.: Quantifying trading behavior in financial markets using Google trends. Sci. Rep. **3**, 1684 (2013)

26. Royston, P.: Algorithm AS 181: the W test for normality. Appl. Stat. **31**, 176–180 (1982)
27. Surowiecki, J.: The Wisdom of Crowds: Why the many are smarter than the few and how collective wisdom shapes business. Economies. Societies and Nation, Little, Brown (2004)
28. Surowiecki, J.: The Wisdom of Crowds. Anchor Books (2005)

Representation of Species Composition

V. Pawlowsky-Glahn, T. Monreal-Pawlowsky and J. J. Egozcue

Abstract The Aitchison geometry of the simplex, the sample space of compositional data, allows statistical modelling and analysis of compositions without the problems derived from spurious correlation. Here, it is used to show that it offers an alternative to the de Finetti ternary diagram for representing variability of species composition avoiding the problems typical of a standard analysis of proportions, namely spurious correlation and limitation to three or at most four components. The method is illustrated with data representing the species composition of Free and FAD tuna school sets sampled in the Indian and Atlantic Oceans during the 2002–2008 period by purse seiners.

Keywords Simplex · Ternary diagram · Aitchison geometry · Tuna · Fisheries

1 Introduction

Difficulties for a visual illustration of the variability of species composition of sampled sets landed by purse seiners in the Indian and Atlantic Oceans moved Fonteneau et al. [14] to propose the use of ternary plots, named after de Finetti [7, 9], to solve this problem. As summarised by Howarth [17], there has been an extensive use of the ternary diagram in many different science fields. For example, ternary plots can be

V. Pawlowsky-Glahn (✉)
Department of Computer Science, Applied Mathematics, and Statistics,
University of Girona, Girona, Spain
e-mail: vera.pawlowsky@udg.edu

T. Monreal-Pawlowsky
IZVG, Keighley, UK
e-mail: t.monreal@izvg.co.uk

J.J. Egozcue
Department of Civil and Environmental Engineering,
Technical University of Catalonia, Barcelona, Spain
e-mail: juan.jose.egozcue@upc.edu

J.A. Martín-Fernández and S. Thió-Henestrosa (eds.), *Compositional Data Analysis*, Springer Proceedings in Mathematics & Statistics 187,
DOI 10.1007/978-3-319-44811-4_11

found in the Geosciences, known in this field as ternary diagrams [26], and they are used to represent the Hardy–Weinberg law of equilibrium in the Biosciences [16].

The ternary diagram has a major drawback in that only three part compositions can be visualised. A similar representation with four parts is possible using a tetrahedron. More difficult it is to visualize compositions with more parts. Nevertheless, the most important drawback is that, using standard statistical methods with proportions is known to lead to spurious, nonsensical results [8, 31]. Thus, it appears desirable to have tools that allow the visualisation and analysis of a larger number of parts. These tools are available in the framework of compositional data analysis based on the logratio approach. Here they are illustrated with real data corresponding to species composition of tuna landings. In addition to this representation of species composition, it is shown how different groups can be compared using an ANOVA approach.

2 The Data

The data were kindly provided by Scientists from the IRD (Institut de Recherche pour le Développement, France—http://www.ird.fr/). They correspond to species composition of tuna landings. It is a subset of the so-called "SPECIES" files, i.e. of the detailed species composition of all the sampled sets collected in the Indian and Atlantic oceans. These data have been used and described in Fonteneau et al. [14] to introduce the de Finetti ternary diagrams to show species composition. They correspond to data collected during the period 2002–2008 on the landing or transshipment operations of the EU and associated purse seiners.

The characteristics of the data are summarised in Table 1. The available data correspond to 5784 operations in the Atlantic Ocean and 4947 in the Indian Ocean, out of which a part has been obtained using a fish aggregating device (FAD), while another corresponds to free schools (BL). For each ocean and each fishing mode three species of tuna have been recorded, YFT = yellowfin, SKJ = skipjack, and BET = bigeye. The data present a major number of zeros, which distribution is summarised in Table 2. Once the two cases with three zeros have been removed, a graphical representation in a ternary diagram is possible, as can be seen in Fig. 1. Zero values appear in the border of the ternary diagram. If a single component is zero, the data point appears on an edge. More precisely, each data point is placed so that

Table 1 Sample size of species composition of tuna landings by ocean and fishing mode

		Fishing mode	
	N	FAD	BL
Atlantic Ocean (AO)	5784	3526	2258
Indian Ocean (IO)	4947	2857	2090

N = total sample size; FAD = fish aggregating device; BL = free schools

Table 2 Number of samples with zero and non-zero counts in species composition of tuna landings (see text for details)

Ocean	AO		IO	
Assoc	FAD	BL	FAD	BL
No zeros	3238	515	2452	98
Zeros only in YFT	15	7	4	2
Zeros only in SKJ	6	428	10	549
Zeros only in BET	237	164	340	147
Zeros in YFT and SKJ	0	0	0	6
Zeros in YFT and BET	22	182	19	310
Zeros in SKJ and BET	8	960	32	976
Zeros in YFT, SKJ and BET	0	2	0	2

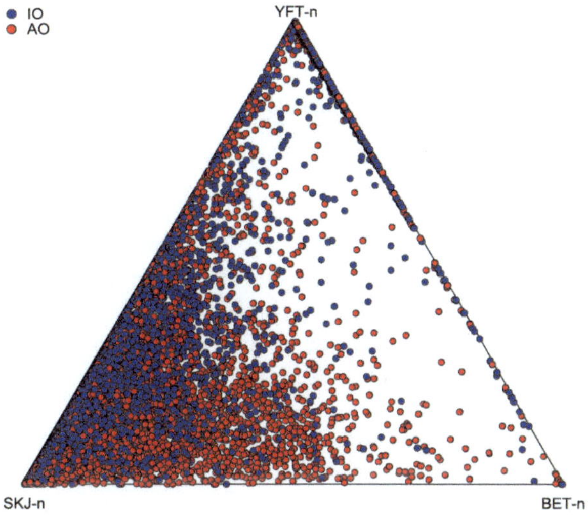

Fig. 1 Representation in the ternary diagram of proportions of landed species. Indian ocean in *blue*, Atlantic ocean in *red*. Data-points on the vertices and edges of the triangle correspond to data with two or one zero-proportions

the edge is partitioned into two segments, according to the proportion between the two non-zero species. When there are zero-proportions in two parts, the data point is represented in the vertex corresponding to the only recorded species, independently of the value of the single component: the proportion of this species is one. However, Fig. 1 may be confusing, as data-points with small proportions in one or two species, appear close or very close to those for which there is a zero proportion. This fact redirects us to a deeper problem concerning the representation of proportions. It is the question of the scale. For a couple of proportions like 0.01 and 0.005, it is clear that the first doubles the second; consequently, their difference or distance should be important. Compare it to the difference between 0.500 and 0.505; in this case, the difference can be considered irrelevant. We say that proportions carry only relative information and that the scale is relative. This relative scale is not shown in a ternary diagram, thus claiming for a change of scale for the representation of proportions. The way out is to use logratio transformations of the data in proportions. They account for the relative scale of compositional data and, at the same time, open up a way for applying standard statistical analysis to logratio transformed data [3, 23, 30].

3 Methods

We base our approach on the fact that compositional data, understood as equivalence classes [6], can be represented as proportions in a simplex; that the simplex is then a representant of the sample space, and that it admits a Euclidean space structure [29], which leads to the so-called *Aitchison geometry*. The approach is based on the use of general scale invariant logratios, also called log-contrasts.

Denoting the abundances of D species by $\mathbf{x} = [x_1, x_2, \ldots, x_D]$, log-contrasts appear as expressions like

$$\ln \frac{x_1^{\alpha_1} x_2^{\alpha_2} \ldots x_k^{\alpha_k}}{x_{k+1}^{\alpha_{k+1}} x_{k+2}^{\alpha_{k+2}} \ldots x_D^{\alpha_D}}, \quad \sum_{i=1}^{k} \alpha_i = \sum_{j=k+1}^{D} \alpha_j ,$$

where x_i can represent proportions of landed species in an arbitrary order, and D is the number of considered species, which in the present example is $D = 3$. An important property of such log-contrasts is that their value does not change when the data in proportions is multiplied by any positive constant, e.g. 100 for obtaining percentages, or by an arbitrary positive number when they are expressed in abundances. This is useful in the case of the reference data, which are not given in proportions, but in some non-homogenised units not adding to a prefixed constant. It does not matter, log-contrasts maintain their values independently of the units in which data are presented; in the present case, only ratios between the abundances of some landed species are relevant. There are special cases of logratios which are useful for the representation of compositional data: (a) simple logratios, like $\ln(x_i/x_j)$, (b) centred logratios, used in the *centered logratio transformation*, which is given by

$$\operatorname{clr}[x_1, x_2, \ldots, x_D] = \left[\ln \frac{x_1}{g(\mathbf{x})}, \ln \frac{x_2}{g(\mathbf{x})}, \ldots, \ln \frac{x_D}{g(\mathbf{x})} \right], \qquad (1)$$

with

$$g(\mathbf{x}) = (x_1 x_2 \ldots x_D)^{1/D} = \left(\prod_{i=1}^{D} x_i \right)^{1/D},$$

and (c) balances, which take the form of a logratio of geometric means of groups of components. In the case of the reference data, with $D = 3$, a possible choice of balances is

$$b_1 = \sqrt{\frac{2}{3}} \ln \frac{x_3}{(x_1 x_2)^{1/2}}, \quad b_2 = \sqrt{\frac{1}{2}} \ln \frac{x_1}{x_2}. \tag{2}$$

They are Cartesian coordinates of the composition of the three species considered.

The use of logratios to represent compositional data impede the use of zero-proportions, since logarithms of zero take negative infinite values. When zero values in abundances or proportions are understood as values below a detection limit, it is advisable to replace the zero values using some imputation criteria [20, 21, 24, 25]. In the present case, with $D = 3$ and a large number of zeros, all types of imputation introduce serious distortions in the data set, as will be shown below.

An important exploratory tool in compositional data analysis is the clr-biplot [4]. It is based on the principal component analysis of compositional data [2] and is a compositional version of the biplot for real data [15]. Elementary explanations can be found in Pawlowsky-Glahn et al. [30] and Thió-Henestrosa et al. [32]. Construction of a clr-biplot starts transforming the data set into their clr's (Eq. 1); a second step is centering the clr-components to place the origin of axes in the center of the data; a singular value decomposition (SVD) of the centered clr-data is then carried out, leading to orthogonal axes which show maximum variability of the data set; finally, the data and the clr-variables are simultaneously represented in a bidimensional plot. The bidimensional projection of the data set provided by the clr-biplot has the advantage that the data can have a large number of parts or components, thus performing a dimension reduction with the minimum loss of variability. In the present case, where $D = 3$, the biplot will represent 100 % of the variability of the data. Biplots can be scaled in several ways, being the so-called *covariance* and *form* biplots the most used. Covariance biplots scale the directions of the clr-variables proportionally to their standard deviations, and the components of data-points appear in a standardised way. Covariance biplots are adequate to visualise relationships between clr-variables. Form biplots normalise the representation of clr-variables, leaving the data-points so that their inter-distances are the Aitchison distances in the simplex [2, 3, 29]; thus, they are adequate to study the distribution of data-points.

To start with the analysis of the reference data set of landed tuna species, the zeros have been substituted all by the value 0.00005 in proportions. This value is arbitrary and is chosen to visualise the effect of this raw replacement in a clr-biplot. Figure 2 shows the covariance biplot where the data-points have been coloured according to the appearance of zeros (Table 2). In the legend, a Z followed by the acronym of a species indicates zeroes in that (or those) species. The biplot shows a clear separation of the groups corresponding to samples with zeros, thus demonstrating that replacement of zeros introduces artefacts in the data set. From a compositional perspective, this zero replacement respects the ratios between the non-zero parts. In the present case, with $D = 3$, the dimension of the data set is 2 and, fixing one logratio, the degrees of freedom of a replacement of one zero is one. This effect is shown in Fig. 2, in which those cases with only one replaced zero appear dispersed and separated from data-points with no-zeros (NZ, blue points), while those cases where the replacement is carried out in two components, substitution has two degrees of freedom and the replaced data-points appear aligned (grey, pink and yellow points). This

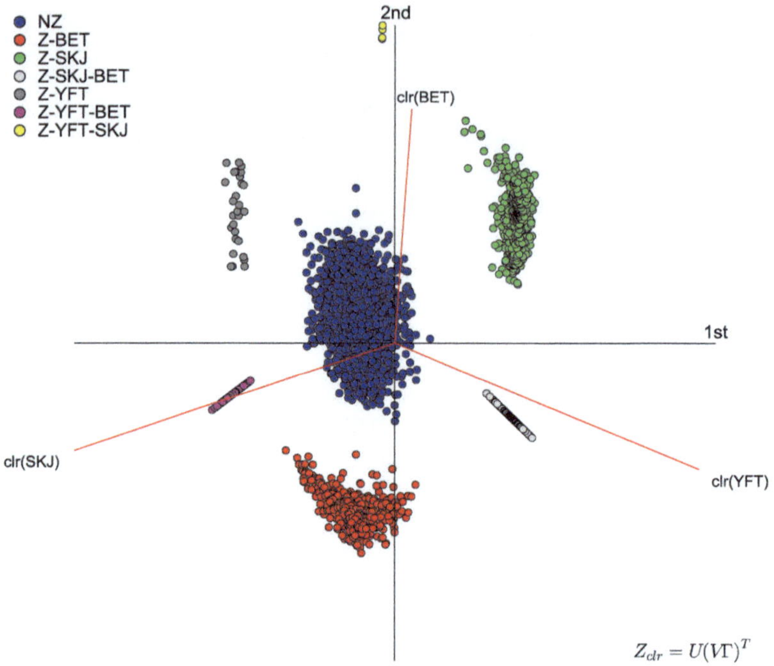

Fig. 2 Biplot of data after zero replacement; grouped by categories with zero counts (see text for details). Explained variability 100 %

kind of artefacts due to zero replacements appear with any procedure of zero replacement in a composition with three parts ($D = 3$); less rough replacement procedures can reduce the separation of the replaced data-points from the data-points without zeros, and even introduce a dispersion similar to that of data without zeros [21].

For the subsequent analysis, data with zeros have been omitted. The whole data set is classified by ocean (Indian IO, Atlantic AO) and by fishing mode (FAD, BL). Under this two-way classification, the biplot is computed assigning colours to the four different groups of data points. Figure 3 shows the form (upper panel) and covariance (lower panel) biplot. The form biplot (upper panel) is adequate to show the position of the data-points. The length of rays is the projection of a unitary vector in the direction of each clr-variable. As in this case $D = 3$, the represented variance is 100 %, and the rays appear of equal length and unitary, thus telling that they are perfectly projected on the plane of the two principal components. In cases of larger dimension, a short ray points out that the clr-direction is not well projected on the plane. Looking at the data, it can be observed that the classes clearly overlap, although a small shift to the left of data from AO–FAD relative to IO–FAD can be observed. Other classes are almost hidden by these two classes. As the first principal component is approximately a logratio of the geometric mean of YFT and SKJ over BET, the shift to the left of AO–FAD could be due to a difference in species composition in both oceans, leading

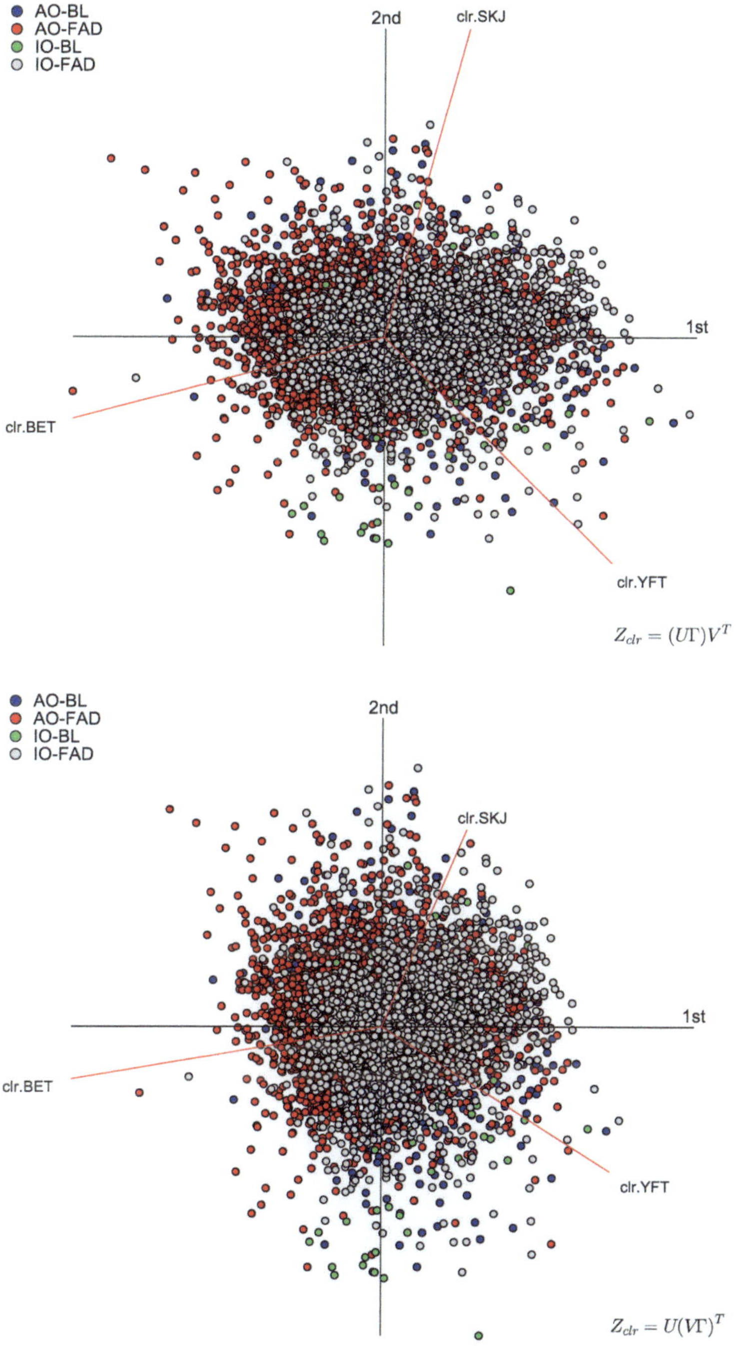

Fig. 3 Biplots of data without zeros. Explained variability 100 %. *Upper panel* form biplot by ocean and fishing mode. *Lower panel* covariance biplot by ocean and fishing mode

to a higher effectiveness of FAD procedures in capturing BET, relative to YFT and SKJ, in the Atlantic Ocean than in the Indian Ocean. However, this is only a visual impression and this kind of conclusions should be confirmed statistically.

The lower panel is a covariance biplot. As $D = 3$, not many differences can be seen in comparison to the form biplot (upper panel). In a covariance biplot attention is focussed on the clr-variables represented as rays from the origin (in red). The length of the rays is proportional to the standard deviation of each clr-variable corresponding to the logratio of each landed species over the geometric mean of the other two. In a covariance biplot, the main tool of interpretation relays on the segments linking the extreme points of the rays, called links for short. The length of a link is proportional to the standard deviation of the simple logratio between the species in the respective labels. The links in the lower panel of Fig. 3 are quite similar, although the link between clr(SKJ) and clr(YFT) is a bit shorter than the link between clr(SKJ) and clr(BET), or the link between clr(YFT) and clr(BET). A very short link would imply association between the two involved parts, suggesting that the proportions between landed species is almost proportional across the sample [10, 18, 19].

Features observed in the covariance biplot of Fig. 3 are quantified in the variation array [3], shown in Table 3. The three variances of simple logratios in the upper triangle are of the same order of magnitude, indicating there is no special association between parts. Their square roots, the standard deviations, are proportional to the links visualised in the covariance biplot of Fig. 3. Mean values of the logratios are shown in the lower triangle of Table 3. For instance, the sample mean of $\ln(\mathrm{BET}/\mathrm{SKJ})$ is -2.34 which indicates that, in overall mean, the landing abundance of BET is small relative to SKJ landings. Variation arrays by ocean, by fishing mode, or by both, show no significantly different features and are not shown here.

In the form biplot (Fig. 3, upper panel), the coordinates of data-points on the first principal axis are approximately proportional to the following log-contrast

$$\ln \frac{(\mathrm{YFT})^{0.58}(\mathrm{SKJ})^{0.21}}{(\mathrm{BET})^{0.79}}, \qquad (3)$$

which corresponds to the projections of rays on the first principal axis. However, the interpretation of such coordinates is not easy, specially when there are more than

Table 3 Variation array of data without zeros

Species	YFT	SKJ	BET
YFT		1.66	3.56
SKJ	1.43		2.62
BET	−0.91	−2.34	

Upper triangle contains variances of logratios of species by row and column. Lower triangle shows the mean of the logratios. Total variance is 2.61

three species in the data set. Therefore, it is common to use more simple log-contrasts to represent and analyse the compositions. This is done using balances, like those in Eq. (2).

Within the Aitchison geometry of the simplex, it is advisable to work on orthonormal coordinates, called isometric logratio (ilr) coordinates [13] to which all the standard statistical methods, devised for real-random variables, can be applied [23]. Those coordinates, shown in Fig. 3 (upper panel) or in Eq. (3), are ilr-coordinates generated in the principal component analysis. However, in practice, the coordinates used correspond to a sequential binary partition, as described in Egozcue et al. [13]; Egozcue and Pawlowsky-Glahn [11, 12] and Pawlowsky-Glahn and Egozcue [28]. The generated ilr-coordinates are then called balances and have a simpler form (Eq. 2). For the reference data, a three part composition, the balance-coordinates used here are

$$b_1 = \sqrt{\frac{2}{3}} \ln \left(\frac{\text{BET}}{(\text{YFT} \cdot \text{SKJ})^{1/2}} \right) , \tag{4}$$

and

$$b_2 = \sqrt{\frac{1}{2}} \ln \left(\frac{\text{YFT}}{\text{SKJ}} \right) . \tag{5}$$

They can be visualised in a dendrogram like the one in Fig. 4. These kind of balance dendrograms [28, 33] are designed to show graphically and simultaneously: (a) the sequential binary partition used for designing the coordinates; (b) the decomposition of the total variance by coordinates; (c) dispersion of balances using box-plots; (d) mean values of balances; and (e) all these features for different populations or groups. In Fig. 4, the horizontal junction from the group (SKJ, YFT) to BET corresponds to the balance b_1 (Eq. 4); the horizontal junction of SKJ and YFT corresponds to the balance between these two species denoted b_2 (Eq. 5). Note that in presence of more species the dendrogram would have as many junctions as the number of parts minus one. The length of vertical bars over the junction of each balance is proportional to the variance of the balance, thus constituting a decomposition of the total variance of the sample. The horizontal junction between groups of parts is used to show the dispersion of the balance. All horizontal junctions represent the same range and are scaled accordingly. The zero of the balance (equality of the numerator and denominator of the balance) is the central point, independently of the length of the junction. The boxplots under the junction visualize the dispersion of the sample balances. In the scale of the horizontal junction, the fulcrum of the vertical bars is the mean of the corresponding sample-balance. This is, when the mean balance is placed at the left side, as is the case in Fig. 4, it points out that the parts on the left have greater proportions than the parts placed at the right: it works like a lever in equilibrium i.e. a balance in the plain sense. The whole structure is built up using the whole sample, which corresponds to the black vertical lines. When the sample is divided into different populations the variance decomposition of each sub-population is superimposed, respecting the scales of the horizontal junctions. In this way, both variances and mean of each balance can be visually compared.

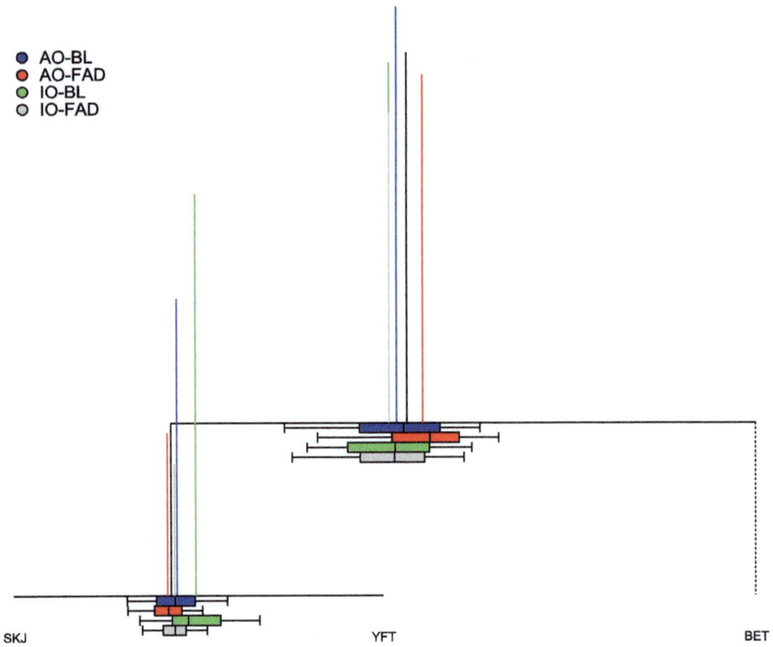

Fig. 4 Dendrogram for data without zeros grouped by ocean and fishing mode

In Fig. 4 the whole sample of landed species (no zeros) is divided by ocean and by association, giving rise to a classification into four groups. As can be seen, variability within each group is pretty similar and some differences might be significant in the means of the balances. For example, the group of tuna landed in the Atlantic Ocean using FAD (FADAO, red) seems to be different in b_1, while the group of tuna landed in the Indian Ocean from BL (BLIO, green) shows a more differentiated mean in b_2. Boxplots for b_1 and b_2 are reproduced with more detail in Fig. 5 (left and right panels respectively).

Assuming the data to be obtained by simple random sampling from a statistical population, one can apply standard statistical methods to balance-coordinates, including model adjustment or testing of hypothesis, e.g. testing equality of means or performing an analysis of variance. Figures 6 and 7 represent the data in ilr-coordinates (Eqs. 4 and 5), balances in this case, after removing samples with zeros. Data have been split into four sets, crossing Atlantic Ocean (AO) and Indian Ocean (IO) with the extraction techniques, FAD and BL. For illustration of our approach, under the assumption of simple random sampling, a bidimensional normal distribution has been adjusted to the data represented by balances. Some isodensity contours of the normal distribution are shown in these figures. Before carrying out a statistical test on the goodness-of-fit, the normal distribution for the balance-coordinates seems a first option for modelling. This corresponds to a normal distribution on the simplex [1, 5, 22, 30]. For the whole sample a low, but significant, correlation coef-

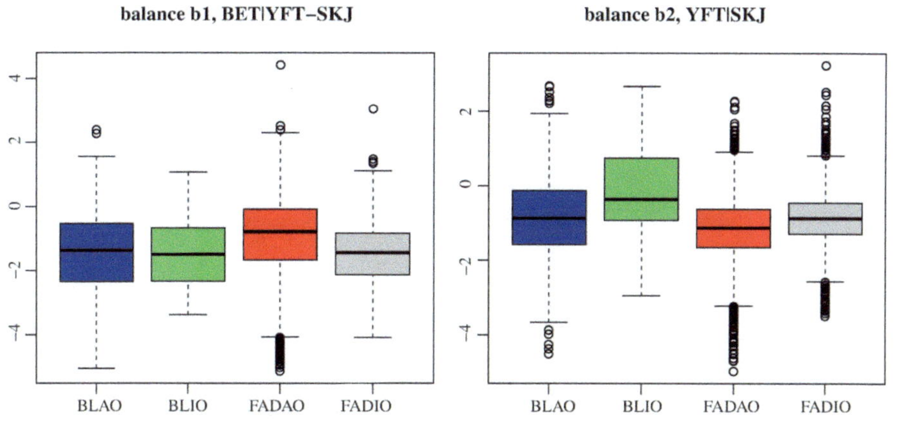

Fig. 5 Boxplots of balance b_1 (*left*) and balance b_2 (*right*) (enlarged)

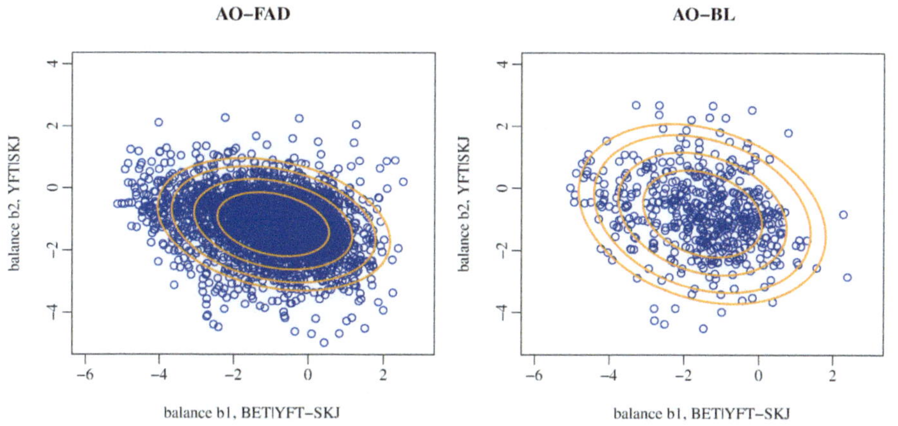

Fig. 6 Atlantic ocean data without zeros represented in balance-coordinates (b_1, b_2). Normal contours (probabilities, 0.5, 0.75, 0.90, 0.95) fitted to data. *Left panel*, FAD-fishing mode; *right panel*, BL-fishing mode

ficient of −0.23 is obtained when comparing b_1 and b_2. Separating the four groups, the correlation coefficients were −0.25 for AO–BL, 0.33 for IO–BL, −0.26 for AO–FAD, and −0.07 for IO–FAD, thus indicating a weak linear dependence between the balances in the four groups.

A test on equality of means of each balance b_1 and b_2 gave p-values $< 10^{-4}$, indicating that the hypothesis of equality of means should be rejected. The groups responsible of these significative differences for mean balances are those mentioned when looking at the dendrogram in Fig. 4 and the boxplots in Fig. 5. For instance, the whiskers of boxplots approximate 95 % confidence intervals on the mean values of the balances. For balance b_2 the only mean differing significantly from the others is that of IO–BL. For balance b_1, the mean differing significatively from the others is

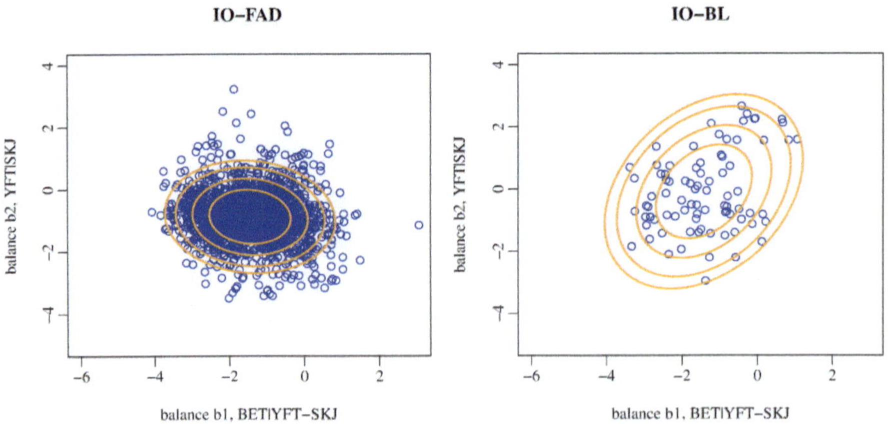

Fig. 7 Indian ocean data without zeros represented in balance-coordinates (b_1, b_2). Normal contours (probabilities, 0.5, 0.75, 0.90, 0.95) fitted to data. *Left panel*, FAD-fishing mode; *right panel*, BL-fishing mode

AO–FAD, specially when compared with IO–BL. This means that the proportions of the three tuna species landed depend on the ocean and on the fishing mode, although the difference in mean is not strong.

4 Conclusions

Representation of a composition of species in ternary diagrams, in this case landed species of tuna, was proposed by Fonteneau et al. [14]. Here, we show that this representation can be complemented using the exploratory and graphical tools of the compositional data analysis. Ternary diagram representations have two main shortcomings: proportions are not represented in an appropriate scale; and they cannot be generalised to more than four species. Logratio analysis overcome these shortcomings. The visual tools proposed are: the compositional biplot, the balance dendrogram, and scatterplots of balance coordinates. Compositional biplots simultaneously represent data in ilr-coordinates (form biplot) and allows a quite intuitive visualisation (covariance biplot) of relationships between variables. Balance dendrograms allow comparison of means and variances of balance-coordinates, selected by the analyst, from different populations. They graphically show all the elements for an ANOVA analysis to compare means of balances. Finally, representation of compositions by ilr-coordinates allow to construct all desired scatterplots. Statistical modelling of balances or other ilr-coordinates is reduced to the standard statistical multivariate techniques. Moreover, other sets of log-contrasts can be studied separately using standard techniques. Finally, all these representations can be generalised to a large number of parts. For a number of parts greater than 3, the clr-biplot is an optimal projection; the balance dendrogram grows as a tree; and the selection of interpretable balance-coordinates gets more involved, but feasible. In summary, working on coordinates of the simplex is a powerful tool for the representation and analysis of compositional data.

Acknowledgments The authors would like to thank two reviewers, A. Laurec and M. Pierotti, for their constructive comments on an earlier version of this paper. The data used for illustration were kindly provided by the authors of Fonteneau et al. [14]. This research has been supported by the *Spanish Ministry of Education and Science* under projects: 'Ingenio Mathematica (i-MATH)' (Ref. No. CSD2006-00032), and 'CODA-RSS' (Ref. MTM2009-13272), from the *Spanish Ministry of Economy and Competitiveness* under the project 'METRICS' (Ref. MTM2012-33236); and from the *Agència de Gestió d'Ajuts Universitaris i de Recerca* of the *Generalitat de Catalunya* under the project Ref. 2009SGR424.

References

1. Aitchison, J.: The statistical analysis of compositional data (with discussion). JRSS Ser. B (Stat. Meth.) **44**(2), 139–177 (1982)
2. Aitchison, J.: Principal component analysis of compositional data. Biometrika **70**(1), 57–65 (1983)
3. Aitchison, J.: The Statistical Analysis of Compositional Data (Reprinted in 2003 by The Blackburn Press), p. 416. Chapman & Hall Ltd., London (UK) (1986)
4. Aitchison, J., Greenacre, M.: Biplots for compositional data. JRSS Ser. C (Appl. Stat.) **51**(4), 375–392 (2002)
5. Aitchison, J., Shen, S.M.: Logistic-normal distributions. Some properties and uses. Biometrika **67**(2), 261–272 (1980)
6. Barceló-Vidal, C., Martín-Fernández, J.A., Pawlowsky-Glahn, V.: Mathematical foundations of compositional data analysis. In: Ross, G. (ed.) Proceedings of IAMG 2001 p. 20. Cancun (Mex) (2001)
7. Cannings, C., Edwards, A.W.F.: Natural selection and the de Finetti diagram. Ann. Hum. Genet. **31**, 421–428 (1968)
8. Chayes, F.: On correlation between variables of constant sum. J. Geophys. Res. **65**(12), 4185–4193 (1960)
9. Edwards, A.W.F.: Foundations of Mathematical Genetics, 2nd edn, p. 121. Cambridge University Press (2000). ISBN-13: 978-0521775441
10. Egozcue, J.J., Lovell, D., Pawlowsky-Glahn, V.: Testing compositional association. In: Hron, K., Filzmoser, P., Templ, M. (eds.) Proceedings of CoDaWork 2013, Vorau (AT) (2013). ISBN: 978-3-200-03103-6. 28–36
11. Egozcue, J.J., Pawlowsky-Glahn, V.: CoDa-dendrogram: a new exploratory tool. In: Mateu-Figueras, G., Barceló-Vidal, C. (eds.) Proceedings of CoDaWork 2005, Girona (E) (2005). ISBN: 84-8458-222-1
12. Egozcue, J.J., Pawlowsky-Glahn, V. Simplicial geometry for compositional data. In: Compositional Data Analysis in the Geosciences: From Theory to Practice, Special Publications 264, pp. 145–159. Geological Society of London (2006)
13. Egozcue, J.J., Pawlowsky-Glahn, V., Mateu-Figueras, G., Barceló-Vidal, C.: Isometric logratio transformations for compositional data analysis. Math. Geol. **35**(3), 279–300 (2003)
14. Fonteneau, A., Chassot, E., Ortega-Garcia, S., Delgado de Molina, A., Bez, N.: On the use of the de Finetti ternary diagrams to show the species composition of free and FAD associated tuna schools in the Atlantic and Indian Oceans. Trop. Tunas **65**(2), 546–555 (2010)

15. Gabriel, K.R.: The biplot—graphic display of matrices with application to principal component analysis. Biometrika **58**(3), 453–467 (1971)
16. Graffelman, J., Camarena, J.: Graphical tests for Hardy-Weinberg equilibrium based on the ternary plot. Hum. Her. **65**, 77–84 (2008). doi:10.1159/000108939
17. Howarth, R.J.: Sources for a history of the ternary diagram. British J. Hist. Sci. **29**(3), 337–356 (1996)
18. Lovell, D., Pawlowsky-Glahn, V., Egozcue, J.J.: Have you got things in proportion? A practical strategy for exploring association in high-dimensional compositions. In: Hron, K., Filzmoser, P., Templ, M. (eds.) Proceedings of CoDaWork 2013, Vorau (AT) (2013). ISBN: 978-3-200-03103-6, 100–110
19. Lovell, D., Pawlowsky-Glahn, V., Egozcue, J.J., Marguerat, S., Bähler, J.: Proportionality: a valid alternative to correlation for relative data. PLoS Comput. Biol. **11**(3), e1004075 (2015). doi:10.1371/journal.pcbi.1004075
20. Martín-Fernández, J.A., Barceló-Vidal, C., Pawlowsky-Glahn, V.: Dealing with zeros and missing values in compositional data sets using nonparametric imputation. Math. Geol. **35**(3), 253–278 (2003)
21. Martín-Fernández, J.A., Palarea, J., Olea, R.: Dealing with zeros. See Pawlowsky-Glahn and Buccianti **2011**, 43–58 (2011)
22. Mateu-Figueras, G., Pawlowsky-Glahn, V., Egozcue, J.J.: The normal distribution in some constrained sample spaces. SORT **37**(1), 29–56 (2013)
23. Mateu-Figueras, G., Pawlowsky-Glahn, V., Egozcue, J.J.: The principle of working on coordinates. See Pawlowsky-Glahn and Buccianti **2011**, 31–42 (2011)
24. Palarea-Albaladejo, J., Martín-Fernández, J.A.: A modified EM alr-algorithm for replacing rounded zeros in compositional data sets. Comp. Geosc. **34**(8), 2233–2251 (2008)
25. Palarea-Albaladejo, J., Martín-Fernández, J.A., Gómez-García, J.A.: Parametric approach for dealing with compositional rounded zeros. Math. Geol. **39**(7), 625–645 (2007)
26. Pawlowsky-Glahn, V., Buccianti, A.: Visualization and modeling of subpopulations of compositional data: statistical methods illustrated by means of geochemical data from fumarolic fluids. Int. J. Earth Sci. (Geol. Rundschau) **91**(2), 357–368 (2002)
27. Pawlowsky-Glahn, V., Buccianti, A. (eds.): Compositional Data Analysis: Theory and Applications, p. 378. Wiley & Sons (2011)
28. Pawlowsky-Glahn, V., Egozcue, J.J.: Exploring compositional data with the CoDa-Dendrogram. Austr. J. Stat. **40**(1 & 2), 103–113 (2011)
29. Pawlowsky-Glahn, V., Egozcue, J.J.: Geometric approach to statistical analysis on the simplex. SERRA **15**(5), 384–398 (2001)
30. Pawlowsky-Glahn, V., Egozcue, J.J., Tolosana-Delgado, R.: Modeling and Analysis of Compositional Data. Wiley & Sons, Chichester UK (2015). 272 pp
31. Pearson, K.: Mathematical contributions to the theory of evolution. On a form of spurious correlation which may arise when indices are used in the measurement of organs. Proc. R. Soc. Lond. *LX*, 489–502 (1897)
32. Thió-Henestrosa, S., Daunis-i-Estadella, J., Barceló-Vidal, C.: Exploratory compositional data analysis using CoDaPack 3D. See Pawlowsky-Glahn and Buccianti **2011**, 329–340 (2011)
33. Thió-Henestrosa, S., Egozcue, J.J., Pawlowsky-Glahn, V., Kovács, L.Ó., Kovács, G.: Balance-dendrogram. A new routine of CoDaPack. Comp. Geosc. **34**(12), 1682–1696 (2008)

Joint Compositional Calibration: An Example for U–Pb Geochronology

R. Tolosana-Delgado, K.G. van den Boogaart, E. Fišerová, K. Hron and I. Dunkl

Abstract This contribution explores several issues arising in the measurement of a (geo)chemical composition with Laser Ablation Inductively Coupled Plasma Mass Spectrometry (LA-ICP-MS), specially in the case that the quantities of interest are linear functions of (log)-ratios. These quantities are scale invariant, but in general cannot be estimated without taking into account possible additive noise effects of the instrumentation, incompatible with a purely compositional approach. The proposed ways to a solution heavily build upon the multi-Poisson distribution, highlighting the counting nature of the readings delivered by these instruments. The model can be fitted using a generalized linear model formalism, and it allows for a joint calibration of all components at once. Relevance of these considerations is shown with some simulation studies and in a real case of multi-isotopic geochronological analyses. Results suggest that the most critical aspect of this analytical technique is the assumption that the amount of ablated mass per second between samples of unknown and known compositions is similar (*matrix matching*): if this cannot be ensured, absolute estimations of the abundance of each of these isotopes fails, while their (log)ratios are perfectly estimable. This opens the door to using the model for a joint calibration by loosening the condition of matrix matching and using several standards of different composition.

Keywords Poisson regression · GLM · Count composition · Multi-element calibration · Concordia plot apologies for the delay

---

R. Tolosana-Delgado (✉) · K.G. van den Boogaart
Helmholtz Zentrum Dresden-Rossendorf, Helmholtz Institute Freiberg for Resource Technology, Freiberg, Germany
e-mail: r.tolosana@hzdr.de

E. Fišerová · K. Hron
Department of Mathematical Analysis and Applications of Mathematics, Palacky University, Olomouc, Czech Republic

I. Dunkl
Department of Sedimentology and Environmental Geology, Georg August University, Goettingen, Germany

J.A. Martín-Fernández and S. Thió-Henestrosa (eds.), *Compositional Data Analysis*, Springer Proceedings in Mathematics & Statistics 187,
DOI 10.1007/978-3-319-44811-4_12

Abbreviations

Concept	Abbreviation (definition)
Indices for variables	i, j, l
Index for time slices	k
Measuring interval	T_k
Background time window	$B_k = [t_A^k, t_b^k]$
Signal time window	$M_k = [t_M^k, t_N^k]$
A time slice from interval T_k	t_k
Reading of variable i at time slice t_n	$x_i(t_n)$
Background level of variable i at interval T_k	b_{ki}
Signal of variable i at t_n	$\Delta x_i(t_n) = x_i(t_n) - b_{ki}$
Signal level of variable l at interval T_k	y_{kl}
Average readings of variable x_i	$\bar{x}_i$
Standard deviation of readings of variable x_i	s_{xi}
Average background associated to variable x_i	b_i
Standard deviation of the background b_i	s_{bi}
Measuring channel	i
Number of channels	P
(Random) reading	X_i
True composition	$\mathbf{Z} = [z_1, z_2, \ldots, z_D]$
Expected background counts per second at channel i	λ_{bi}
i-th channel dwell time duration	ω_{0i}
Sensitivity of channel i to isotope j	λ_{ij}
Matrix of sensitivities	$\mathbf{\Lambda}$
Total number of counts produced by isotope j	λ_j
Proportion of counts in each channel i produced by isotope j	Λ_{ij}^*
Number of analytes analysed in this session	K
Index for one analyte	k
Set of indices for standard analytes	$\mathcal{K}_s$
Set of indices for sample analytes	$\mathcal{K}_m$
Total set of analyte indices	$\mathcal{K} = \mathcal{K}_s \cup \mathcal{K}_m$
Composition-to-counts model	$\mathbf{\Lambda}(\cdots; t)$
Vector of expected background counts	$\boldsymbol{\lambda}_b$
Other parameters of the model	$\boldsymbol{\theta}$
Whole measurement period	$T = \bigcup_{k=1}^{K} T_k$

Concept	Abbreviation (definition)
True composition of sample k	$\mathbf{Z}_k$
Nominal composition, if sample k is a standard	$\mathbf{z}_k$
Union of background windows of the session	$B = \bigcup_k B_k$
Calibration data set	$\mathbf{X}^s$
Prediction data set	$\mathbf{X}^m$

1 Introduction

Laser Ablation Inductively Coupled Plasma Mass Spectrometry (LA-ICP-MS) is an in-situ analytical technique to quantify the abundance of some isotopes (or their ratios) in a sample of unknown composition. The matter to analyse is ablated with a laser and the resulting aerosol is introduced in a plasma, where its atoms' part with their most loosely bound electron to form a (+1)-charged ion. These are then separated by an electromagnetic field, on the basis of their mass-charge ratio, each colliding with a detector. The result is a vector of counts (number of detected ions per unit of time) for several masses, which must then somehow be related to concentrations or relative abundances of the several isotopes. Usually unknowns (called "samples") and materials with known compositions (standard reference materials, later shortly as "standards") are used alternately. This common procedure is called as unknown-standard bracketing and performed in order to derive the proportionality factor between signal and composition. The detectors receive counts even when no ablation happens, forming what is often called a *background* or *blank*. Moreover, the signal received while analysing standards shows systematic drifts at several time scales (along the day, during a measurement, etc.), even though the standards have homogeneous compositions. Thus, the desired ratios of abundances must be estimated from data on counts, taking into account the noise in the signal and a proportionality between ratios of signals and ratios of abundances along time, which is in general non-constant. This problem is typically dealt with by establishing a calibration line for each element; these are obtained by subtracting an additive noise to the signal and fitting some parametric function of time to the proportionality values obtained from dividing the readings for the standards by their normative value. Each isotope is calibrated separately. Several functions have been reported to be used, though mostly they are piecewise linear segments or polynomial fits [2]. It is important to mention that conventional calibration procedure of LA-ICP-MS instruments is targetted to obtaining estimates of the amount of each element *individually* with the best precision and accuracy possible, i.e. unbiased and with lowest variability in absolute terms. However, these values are most often used to compute some informative (log)ratios.

None of the existing methods considers in any sense the possible compositional nature of the problem [7], namely the fact that the target vector to estimate is a composition, which introduces some modifications on the setting. Two definitions exist of a composition, each with its own implications. The classical definition states that a composition is a vector of positive components and a constant sum to 100 %. This definition implies that results of the measurement procedure should deliver vectors of amounts on all elements of the periodic system which are always non-negative and sum to 100 %. Because of the lack of a joint calibration/measurement model, one does not know whether this constant sum will be honored, or how can results be corrected to satisfy it. The modern definition states that a (chemical) composition is a vector of positive components reporting the relative abundance of (the set of all) elements in a sample, or alternatively, which total sum is irrelevant (a property called *scale invariance*). This definition begs the question of why should we spend efforts in obtaining *individually* unbiased and lowest variability estimates of each of these quantities in absolute terms, if we are going to interpret them *jointly* in terms of a relative scale.

This contribution presents several models and methods to take the compositional nature of the problem into account, in its two definitions and with all implications outlined before. The paper discusses and shows the potential uses and limitations of each of these models, compared with the classical approach. This work builds upon materials on from Fišerová et al. [3] and the considerations regarding the background noise in measuring geochemical compositions by van den Boogaart et al. [10]. The keystone of our approach is given by the multi-Poisson distribution [11]. This distribution is chosen because it is the most parsimonious model able to describe the number of counts observed in a series of categories when the total number of counts is not known. The multi-Poisson distribution is an extremely simple model, with many shortcomings (e.g., there is no way to model the dependence between the counts in two categories; or the fact that the variance and the mean must be equal for each component). However, its simplicity makes it a very good model when one does not have enough information to model the actual physics of the phenomenon.

The paper is distributed as follows. Section 2 presents the fundamentals of the LA-ICP-MS analytical technique and its current practice for the non-expert readers. Section 3 puts forward two stochastic models of generation of LA-ICP-MS signals that include both compositional considerations and simplified physics, and gives reasons to the choice of the multi-Poisson model. Section 4 presents a statistical method to work with each of these models, in both the calibration and estimation settings. Section 5 uses a series of simulated scenarios to show the potentials and limitations of these models, specially with regard to their ability to produce unbiased estimates on absolute or relative terms under distortion from the model hypotheses. Section 6 shows the usefulness of one of these models in a real case study. Finally Sect. 7 discusses the main aspects raised by the simulation and real case studies, as a form of preliminary conclusions. Two appendices are included: one summarizing the several geometries involved, and one presenting other compositional calibration models not fitting to the data but included for the sake of completeness.

2 Basics of Laser Ablation Inductively Coupled Plasma Mass Spectrometry (LA-ICP-MS)

2.1 Review of Poisson Distribution Properties

A random variable X is said to follow a Poisson distribution with intensity parameter λ, denoted as $X \sim \mathcal{P}o(\lambda)$ if and only if its probability mass function is

$$f_X(x) = \frac{\lambda^x}{x!} e^{-\lambda}, \quad x = 0, 1, 2, 3, \ldots.$$

The expected value and population variance of X are $\mathrm{E}[X] = \mathrm{Var}[X] = \lambda$, hence this parameter is often called expected number of counts. Note that, in spite of this identification, X is an integer, while the parameter and these statistics are real positive values $\lambda \in \mathbb{R}_+$. Its dispersion coefficient is defined as $D = \mathrm{Var}[X]/\mathrm{E}[X] = 1$ always. If $D > 1$ $(D < 1)$, there is evidence of overdispersion (underdispersion) which implies that there is more (less) variability around the model's fitted values than is consistent with a Poisson distribution. Note that this is not the coefficient of variation.

If λ is large, the Poisson distribution can be excellently approximated by a normal distribution with mean and variance both equal to λ.

The sum of two independent Poisson distributed variates $X_1 \sim \mathcal{P}o(\lambda_1)$ and $X_2 \sim \mathcal{P}o(\lambda_2)$ follows also a Poisson distribution $Y = X_1 + X_2 \sim \mathcal{P}o(\lambda_1 + \lambda_2)$. The difference $Z = X_1 - X_2$ follows a Skellam distribution [8], but if $\lambda_1 \gg \lambda_2$, then $Y = X_1 - X_2 \sim \mathcal{P}o(\lambda_1 - \lambda_2)$ approximately.

A vector of D Poisson variates $X_i \sim \mathcal{P}o(\lambda_i)$ follows a multi-Poisson distribution (van den Boogaart and Tolosana-Delgado 2013, p. 64) with parameter vector $\boldsymbol{\lambda} = [\lambda_1, \lambda_2, \ldots, \lambda_D]$. This is a vector of non-negative integer components characterized by the following conditional construction:

1. the sum of these components $X^T = \sum_i^D X_i \sim \mathcal{P}o(\lambda^T = \sum_i^D \lambda_i)$ gives a total number of counts; note that this holds because the several components are independent by the nature of the Poisson process;
2. conditional on a fixed total number of counts x^T, the number of counts on each category follows a multinomial distribution with parameters $\mathbf{p} = \mathcal{C}[\boldsymbol{\lambda}]$ and $n = x^T$. Here $\mathcal{C}[\boldsymbol{\lambda}]$ means the closure of $\boldsymbol{\lambda}$.

This construction allows to study the multi-Poisson distributions as the product of a (classical univariate) Poisson distribution times a multinomial distribution.

2.2 Technical Procedure

A laser ablation inductively coupled plasma mass spectrometer (LA-ICP-MS) is an in situ analytical instrument that is used for determining the elemental and isotopic composition of micrometer-sized areas of solid materials such as minerals, glasses, metal alloys, bones, teeth, calcareous shell and wood, and fluids trapped as inclusions within solids. The analytes are polished and cleaned before introduction into the laser cell. Typically an excimer UV laser (193 nm) is used for ablation, the generated aerosol from the approximately 20 to 100 μm diameter laser spot is transported by a helion–argon gas mixture into the ICP-MS. Here the particles of the aerosol are dried, atomized, and partly ionized by the plasma. The resulting positive ions are accelerated and separated in variable electric and magnetic fields and the ions are distinguished according to their mass-to-charge ratio.

One or more detectors is counting the ions at the end of the trajectories during a certain time period (*dwell time*). Then the detectors move to different positions, corresponding to other masses, and count impacts there. And so on, until the complete set of masses has been visited. Afterwards the detectors come back to the starting position and counts again, thus starting a new time slice. The dwell time can be variable for the different analytes according to their abundance or importance Several of such time slices occur during a second, producing a multivariate reading. In spite of the sequential character of the measurements within a time slice, these are considered simultaneous.

This reading procedure is applied in three different situations:

background the readings obtained with no material,
standard reference material the readings obtained with an analyte of known composition (or short, *standard*),
sample (*sensu stricto*) with an analyte of unknown composition.

Given that the counts are registered for each analytes during a fraction of a second, the resulting total counts per time window are linearly upscaled to counts per second. The analytical sequences are usually organized as *bracketing*, i.e. samples and standards are measured alternately.

2.3 Conventional Data Analysis Approach

Typically, each measuring interval T_k contains a period of background readings and a period of sample or standard readings, with more or less sharp transitions between them. The first step is the identification of time windows of background $[t_A^k, t_B^k] \subset T_k$ and of signal $[t_M^k, t_N^k] \subset T_k$ (see Fig. 8 later on, for an example), a task that is often manually done by the lab analyst guided by some homogeneity statistics.

The second step is the characterization of the background. This is usually done for each measuring period and for each isotope i separately. All readings in the background window, denoted $\{x_i(t_A^k), x_i(t_{A+1}^k), \ldots, x_i(t_B^k)\}$, are considered as independent realizations of a Poisson distribution of unknown parameter λ_i. With the standard assumptions of statistics of Poisson variates, this parameter can be estimated as the mean readings of that variable in the background window, denoted as b_{ki}. Some labs implement a quality control assessment on the dispersion coefficient within the background window

$$\hat{D}_k = \frac{\mathrm{Var}[x_i(t_A^k), x_i(t_{A+1}^k), \ldots, x_i(t_B^k)]}{\mathrm{E}[x_i(t_A^k), x_i(t_{A+1}^k), \ldots, x_i(t_B^k)]} = \frac{s_{bi}^2(T_k)}{b_{ki}}$$

with heuristic rules that suggest a too-strong non-Poissonal regime if the hypothesis $D = 1$ is not acceptable. In this case, typically the analyst reconsiders the choice of background window.

The third step is the definition of the expected readings for each signal window. This is sometimes applied to the absolute readings, sometimes to ratios between two readings. In particular, in U/Pb-geochronological studies the following ratios of interest are commonly used: Pb^{206}/U^{238}, Pb^{207}/U^{235}, Pb^{207}/Pb^{206} and eventually Pb^{208}/Th^{232}. A first approach would be to neglect fractionation effects on those ratios and apply the same procedure of the background at the set of readings $\{\Delta x_i(t_M^k), \Delta x_i(t_{M+1}^k), \ldots, \Delta x_i(t_N^k)\}$ versus $\{\Delta x_j(t_M^k), \Delta x_j(t_{M+1}^k), \ldots, \Delta x_j(t_N^k)\}$. Note that these values are obtained by subtracting the background levels b_{ki} and b_{kj} to the read counts. Some labs work with the arithmetic mean of the ratios $\Delta x_i(t_m)/\Delta x_j(t_m)$, while other work with the ratios of the count means $\overline{\Delta x_i}/\overline{\Delta x_j}$ within the signal window [4]. More elaborate approaches consider the fractionation trend as a line or as a curve, and attempt several ways of extracting a representative average ratio. For instance, an option is to fit a linear regression trend to the fractionation drift and extrapolate it to the time when the laser beam hit the sample. Whichever method is used, at the k-th measurement interval T_k one has a background value b_{kl} and a signal value for each quantity of interest l (ratio or concentration) y_{kl}.

The fourth step is to study the several measurements $\{y_{kl}\}$ available for the standards, which, due to their homogeneity, should be "equal", i.e. ideally realizations of the same random variable. If this can be assumed, then the average of all standard readings y_l^{std} is compared with the known nominal value μ_l^{std}, and all measurements for unknown analytes are upscaled conveniently as

$$y_{kl} = \frac{\mu_l^{std}}{y_l^{std}} y_{kl}.$$

Note that this equation has only sense if one can assume that the sample and the standard behave in the same way during ablation, i.e. that the same amount of mass per second has been ablated and sent to the mass spectrometre. To ensure that, it is common to select standards of the same kind than the sample, something called *matrix matching* [7].

If the several readings of the standard show too obvious systematic drifts, then it is common to assume some functional model for this drift, fit it to the available measurements of the standard, predict its value $y_l^{std}(T_k)$ for a measuring interval T_k and then upscale the value at that interval accordingly

$$y_{kl} = \frac{\mu_l^{std}}{y_l^{std}(T_k)} y_{kl}.$$

Some effort in these data processing steps is devoted to evaluating the "measurement error" (statistical error, uncertainty). Denoting by $\bar{x}_i$, s_{xi}^2 and b_i, s_{bi}^2 the means and variances of the signal (no background correction) and of the background (dropping the dependence on T_k for simplicity) of the isotope i, the variance of the corrected signal is

$$s_{ci}^2 = s_{bi}^2 + s_{xi}^2,$$

under the hypothesis that the background and the corrected signal are considered as independent Poisson distributions which sums to the uncorrected signal. These concepts and statistics allow as well the definition of *detection limit*, also known as *level of detection* (LoD). The detection limit is defined as that level of signal which cannot be distinguished from the background. It is customary to take the detection limit of each isotope counts as $LoD_i = b_i + 3 \cdot s_{bi}$.

3 A Compositional Calibration Model for ICP-MS

To build a compositional calibration model for this kind of measurements it is important to distinguish between the measurements, the several estimates of some abundances of components of interest, and their actual abundances. In what follows, we call a *channel* a particular position of the detector, which is (hopefully) placed at the end of the trajectory taken by ions of one single charge/mass ratio. P denotes the total number of channels available. The number of collisions counted by the detector on the i-th channel is called a *reading*, and is denoted as X_i. The actual vector of abundances of all D chemical elements in the sample is called its (*actual*) *composition*, and is denoted by the vector $\mathbf{Z} = [Z_1, Z_2, \ldots, Z_D]$.

Following the ideas of the preceding chapter, we may assume the readings of the background to be a vector of P components following a multi-Poisson distribution, i.e. at each time slice while the sample is not being ablated,

$$X_i(t) \sim \mathcal{P}o(\omega_{0i}\lambda_{bi}), \quad t \in [t_A, t_B], \tag{1}$$

where λ_{bi} is the expected counts per second of the background and ω_{0i} is the length of the dwell time of channel i, in seconds. At the moment that ablation starts, though, that channel will produce readings assumed to follow the law

$$X_i(t) \sim \mathcal{P}o\left(\omega_{0i}\left[\lambda_{bi} + \sum_{j=1}^{D} \Lambda_{ij}(t)\cdot(\dot{m}(t)\cdot z_j)\right]\right), \quad t \in [t_M, t_N], \tag{2}$$

where $\dot{m}(t)$ is the mass of sample per second escaping the spot, and $\Lambda_{ij}(t)$ is the expected number of counts per second produced by one gram of element j in channel i. The sample composition is considered constant. The escaped mass per second is often observed to be an exponentially decaying function of time, $\dot{m}(t) = \dot{m}_0 \cdot \exp(\theta_t t)$. This model is called a *full interaction model*, because it allows that each isotope potential influences all channels, in a varying way along time.

The quantities $\Lambda_{ij}(t)$ deserve a longer discussion. Each can be interpreted as a *sensitivity of channel i to element j*. It is typically assumed that each channel i corresponds to one single element j, i.e. that this matrix is diagonal. In this case, $P = D$ and $\Lambda_{ij}(t) := \lambda_j(t)\delta_{ij}$, thus

$$X_i(t) \sim \mathcal{P}o\left(\omega_{0i}\left[\lambda_{bi} + \dot{m}(t)\cdot\lambda_i(t)\cdot z_i\right]\right), \quad t \in [t_M, t_N]. \tag{3}$$

Note that Eq. (3) implies that the process is partly non-compositional, as the intensity of the Poisson process is not scale invariant. This effect can be seen later in the application with true data, in Sect. 6. The expected vector of counts is

$$\mathrm{E}[\mathbf{X}(t)] = \boldsymbol{\lambda}_b + \omega_0 \cdot \dot{m}(t)\cdot[\lambda_1(t)\cdot z_1, \lambda_2(t)\cdot z_2, \ldots, \lambda_D(t)\cdot z_D]$$

which is rather an object of $\mathbb{R}_+^D$, the multivariate positive real space. The part produced by the ablated mass $\dot{m}(t)$ is related to the vector $\boldsymbol{\lambda}(t) \oplus_+ \mathbf{z} = [\lambda_1(t)\cdot z_1, \lambda_2(t)\cdot z_2, \ldots, \lambda_D(t)\cdot z_D]$, where $\oplus_+$ denotes the component-wise product of two vectors of positive components. This operation is the Abelian group operation of $\mathbb{R}_+^D$ [6], also known as amount-perturbation [12]. Thus, this model will be further referred to as an *amount-perturbation upscaling model.* Note that $\boldsymbol{\lambda}(t) = [\lambda_1(t), \lambda_2(t), \ldots, \lambda_D(t)]$ is an *amount vector* of sensitivities. As functions of time, these sensitivities are reported to show very complex and varying patterns at different time scales [2].

Between the full interaction model (Eq. 2) and the amount-perturbation upscaling model (Eq. 3), an intermediate model can be considered. Here we consider the total sensitivity of all channels to ions of type j as varying along time, denoted as $\lambda_j(t)$; however, the way these counts are split among the P channels is considered time-independent, and denoted as Λ_{ij}^*. Thus, $\Lambda_{ij}(t) = \Lambda_{ij}^* \cdot \lambda_j(t)$. The resulting model

$$X_i(t) \sim \mathcal{P}o\left(\omega_{0i}\left[\lambda_{bi} + \sum_{j=1}^{D} \Lambda_{ij}^*\cdot\lambda_j(t)\cdot(\dot{m}(t)\cdot z_j)\right]\right), \quad t \in [t_M, t_N], \tag{4}$$

is called (*constant*) *matrix-interaction amount-perturbation model.*

4 Methods

4.1 Notation and Common Assumptions

Let us assume one particular model $\mathbf{\Lambda}(\mathbf{z}(t_i), \boldsymbol{\lambda}_b, \boldsymbol{\theta}; t_i)$ from those mentioned before, with $\boldsymbol{\lambda}_b$ denoting expected counts from the background level and $\boldsymbol{\theta}$ including the rest of model parameters (dwell times ω_{0i}, sensitivities λ_i, interactions λ_{ij}, eventually including their own trend parameters). A set of readings of count vectors $\{\mathbf{x}(t_i), t_i \in T\}$ is available, obtained along a session T split in K intervals $T_1, T_2, \ldots, T_K$. Each interval contains two non-overlapping windows, the background window $B_k \subset T_k$ and the measurement window $M_k \subset T_k$. All readings during the background window are obtained with no material being analysed, i.e. after Eq. (1). During each measuring window M_k an analyte of different composition was analysed, i.e. $\mathbf{Z}(t_i) = \mathbf{Z}_k$. For some of these compositions, very good estimates $\mathbf{z}_k$ are available (those corresponding to standards): each $\mathbf{z}_k$ is called a *nominal composition*, to distinguish them from the true composition $\mathbf{Z}_k$. Some other of these compositions are totally unknown, and they constitute the actual target of this problem (the samples). Let the set of indices $\mathcal{K} = \{1, 2, \ldots, K\}$ be partitioned in two disjoint subsets $\mathcal{K}_s$ (corresponding to the time intervals when a standard was analysed) and $\mathcal{K}_m$ (corresponding to the intervals when a sample of unknown composition was analysed). The goal is thus to estimate all $\mathbf{z}_k$ for $k \in \mathcal{K}_m$, given the set of all observations $\{\mathbf{x}(t_i), t_i \in T\}$ and the nominal composition of the standards $\mathbf{z}_k$ for $k \in \mathcal{K}_s$. The set of data can also be split in measurements corresponding to all background periods $B = \bigcup_k B_k$ and measurement of standards, the calibration set $\mathbf{X}^s = \{\mathbf{x}(t_i), t_i \in B \cup \bigcup_{k \in \mathcal{K}_s} M_k\}$; and data correspond to measurement windows of samples of unknown composition, the prediction set $\mathbf{X}^m = \{\mathbf{x}(t_i), t_i \in \bigcup_{k \in \mathcal{K}_m} M_k\}$. Given this setting, we will follow the multi-Poisson assumption for the data $\{\mathbf{x}(t_i), t_i \in T\}$, with intensity model $\mathbf{\Lambda}(\mathbf{Z}(t_i), \boldsymbol{\lambda}_b, \boldsymbol{\theta}; t_i)$.

4.2 Generalized Linear Model (GLM)

To apply the formalism of generalized linear models [5] we need to further assume that the composition of the standards is perfectly known and homogeneous, i.e. that the nominal and actual values of the standards are the same, $\mathbf{Z}_k = \mathbf{z}_k$. In this case we can consider the problem divided in two steps:

1. Calibration phase. In this phase we consider only the data available from all background windows and the measurement windows of the standard.
2. Prediction phase. In this phase the model is used to estimate the composition of the unknown samples, potentially with confidence intervals (instead of the classical error). As extra results of this step, one can derive the largest component z_{ki} which cannot be distinguished from 0 with a 99 % confidence interval (the DL for variable i at observation k).

4.2.1 Amount-Perturbation Model with Linear Time Drift

This case is the easiest to understand, as the lack of any form of interaction allows to estimate each component separately. We first reparametrize Eq. (3) using $\dot{m}(t)\lambda_j(t) := \theta_{0j} + \theta_j t_i$, i.e. assumed a linear function of time. In this case we have

$$X_j^s(t_i) \sim \mathcal{P}o(\omega_{0j}[(\theta_{0j} + \theta_j t_i)z_j(t_i) + \lambda_{bj}]).$$

To fit the parameters of this model, the GLM formalism establishes that a transformation $\eta(\cdot)$ of the expected value of $X_j^s(t_i)$ should be predicted with a linear combination of explanatory variables, i.e. in the calibration phase, given that $z_j(t_i)$ and t_i are known everywhere:

$$\eta(\mathrm{E}[X_j^s(t_i)]) = \eta(\omega_{0j}[(\theta_{0j} + \theta_j t_i)z_j(t_i) + \lambda_{bj}]) = a_j + b_j z_j(t_i) + c_j z_j(t_i)t_i.$$

This model can be readily solved if we choose an identity link, i.e. $\eta(x) = x$. This is a non-canonical choice, but allows to trivially identify $a_j = \omega_{0j}\lambda_{bj}$, $b_j = \omega_{0j}\theta_{0j}$ and $c_j = \omega_{0j}\theta_j$ as the statistical versus physical parameters to be estimated. Moreover, in the prediction phase, with estimates $\tilde{a}_j$, $\tilde{b}_j$, $\tilde{c}_j$ set, the unknown is $z_j(t_i)$, i.e. the model becomes again a generalized linear model

$$X_j^m(t_i) \sim \mathcal{P}o(\tilde{a}_j + [\tilde{b}_j + \tilde{c}_j t_i]z_j(t_i)),$$

where $\tilde{a}_j$ is an offset and the predictor variable $[\tilde{b}_j + \tilde{c}_j t_i]$ is known everywhere. This model does not allow an intercept. These models can be estimated in both steps with the GLM maximization likelihood procedures [5].

Note that the canonical choice of the Poisson family (the logarithmic link, $\eta(x) = \log(x)$) would give rise to a different model, called *multiplicative perturbation-scaling model*, and explained in the appendix. It would easily allow the inclusion of the exponential decay observed in $\dot{m}(t)$, but then there would be no way to identify the additive effect of the background with any parameter of the model.

4.2.2 Matrix-Interaction Amount-Perturbation Model with Linear Time Drift

In this case, given the interaction between components it is not possible to consider them totally independently any more. Considering Eq. (2) with $\dot{m}(t)\boldsymbol{\Lambda}(t) =: \boldsymbol{\Theta}^0 + \boldsymbol{\Theta}t_i$, the joint model is

$$\begin{aligned}\mathbf{X}(t_i) &\sim \mathcal{P}o(\boldsymbol{\omega}_0 \oplus_+ [(\boldsymbol{\Theta}^0 + \boldsymbol{\Theta}t_i)\oplus_+ \mathbf{z}(t_i) + \boldsymbol{\lambda}_b]) \\ &= \mathcal{P}o(\mathrm{diag}\,[\boldsymbol{\omega}_0] \cdot [(\boldsymbol{\Theta}^0 + \boldsymbol{\Theta}t_i) \cdot \mathbf{z}(t_i) + \boldsymbol{\lambda}_b]),\end{aligned}$$

or distributing

$$\begin{aligned}\mathbf{X}(t_i) &\sim \mathcal{P}o\left(\text{diag}\left[\boldsymbol{\omega}_0\right]\cdot\boldsymbol{\Theta}^0\cdot\mathbf{z}(t_i)+\text{diag}\left[\boldsymbol{\omega}_0\right]\cdot\boldsymbol{\Theta}\cdot\left[t_i\mathbf{z}(t_i)\right]+\text{diag}\left[\boldsymbol{\omega}_0\right]\cdot\boldsymbol{\lambda}_b\right)\\ &\sim \mathcal{P}o\left(\mathbf{a}+\mathbf{B}\cdot\mathbf{z}(t_i)+\mathbf{C}\cdot\left[t_i\mathbf{z}(t_i)\right]\right).\end{aligned}$$

with intercept vector $\mathbf{a} = \text{diag}\left[\boldsymbol{\omega}_0\right]\cdot\boldsymbol{\lambda}_b = \boldsymbol{\omega}_0 \oplus \boldsymbol{\lambda}_b$ and two matrices $\mathbf{B} = \text{diag}\left[\boldsymbol{\omega}_0\right]\cdot\boldsymbol{\Theta}^0$ and $\mathbf{C} = \text{diag}\left[\boldsymbol{\omega}_0\right]\cdot\boldsymbol{\Theta}$, respectively a K-amounts vector and two $(K \times D)$-matrices of coefficients. Arrived at this point, it is possible to split the problem in the several components of the response, i.e.

$$X_j(t_i) \sim \mathcal{P}o\left(a_j + \mathbf{b}_j' \cdot \mathbf{z}(t_i) + \mathbf{c}_j' \cdot \left[t_i\mathbf{z}(t_i)\right]\right).$$

where $\mathbf{a} = [a_1, a_2, \ldots, a_D]$ and where $\mathbf{b}_j'$ and $\mathbf{c}_j'$ are the j-th rows of the matrices $\mathbf{B}$ and $\mathbf{C}$ respectively. They can be then arranged together in one single vector of counts obtained on different channels, and a single GLM with a response of the Poisson family and an identity link fitted. In this way, one single vector model is fitted,

$$\mathbf{X}(t_i) \sim \mathcal{P}o\left(\mathbf{a}+\mathbf{B}\cdot\mathbf{z}(t_i)+\mathbf{C}\cdot\left[t_i\mathbf{z}(t_i)\right]\right).$$

As we did in the preceding case, in the calibration step we use the data $\mathbf{X}^s$ and the known predictors $\{t_i \in T\}$ and $\{\mathbf{z}_k, k \in \mathcal{K}_s\}$ to obtain estimates $\tilde{\mathbf{a}}$, $\tilde{\mathbf{B}}$ and $\tilde{\mathbf{C}}$ of the model parameters. In the prediction step, by fixing these parameters on their estimated values, the model

$$\mathbf{X}(t_i) \sim \mathcal{P}o\left(\tilde{\mathbf{a}} + \mathbf{Z}(t_i)\cdot[\tilde{\mathbf{B}} + t_i\tilde{\mathbf{C}}]\right)$$

can be fitted with an offset $\tilde{\mathbf{a}}$ and known predictors $[\tilde{\mathbf{B}} + t_i\tilde{\mathbf{C}}]$ to obtain estimates of the coefficient vector $\mathbf{Z}(t_i)$. Note that all channels of $\mathbf{X}$ are considered jointly.

4.3 Caveats

All methods mentioned before share several main limitations:

- perfectly known standard compositions are required;
- it is not possible to model inhomogeneities of the materials considered (standards or samples);
- no solution exists for other more realistic and flexible models, like a multiplicative interaction-perturbation model with additive error;
- if the hypothesis of Poisson distribution is verified to be inappropriate, the likelihood cannot be computed exactly, and GLM fitting procedures might fail.

All these issues can be tackled with Bayesian estimation techniques. Though they are not much more complex than the methods presented so far, these fall beyond the scope of this contribution and are left for future research. Interested readers can consider the work of van den Boogaart et al. [13], dealing with a model that can accommodate some of these effects.

5 Simulation

This section uses simulations from the several models and methods specified before to illustrate their ranges of validity, as well as to show their behaviour when the underlying hypotheses are not satisfied. In particular, it will be shown that estimation of (log)ratios of components is more stable than estimation of the absolute values of these components.

For this task, several scenarios follow. In each scenario, one parameter is left to vary, and for each value of this parameter, we simulate 300 sets of 200 readings of a three-channel instrument ($P = 3$) depending on a four-part composition ($D = 4$), split in $B = 100$ background window readings and $M = 100$ measurement window readings. Only $K = 2$ periods are considered, one for the standard and one for the measurement. The standard is considered to have a nominal composition of $\mathbf{z} = [1, 2, 3, 94]\,\%$, while the sample has an unknown real composition of $\mathbf{z} = [3, 2, 1, 94]\,\%$. As general settings, we consider a sensitivity of $\lambda_i = 2$ counts per unit of mass, a background of $\lambda_i = 2$ counts per dwell time, and a mass of 1000 units ablated. Only results of the three first components are reported. The following cases are considered: varying length of the measurement window; varying total sensitivities and background levels; different masses ablated per unit of time between standard and sample; a heterogeneous standard composition; the presence of a (non-modelled) fractionation effect. These scenarios are built with an amount-perturbation upscaling model. Finally, a constant amount-matrix upscaling-interaction model is considered. In each case, we show boxplots of the (absolute) enrichment/depletion factors z_i^m/z_i^s, i.e. the number with which one should multiply the nominal value of component i on the standard to obtain an estimate of the absolute abundance of that element on the sample of unknown composition. If results are not satisfactory, we also report results of the corresponding perturbation, $\mathcal{C}[z_1^m/z_1^s, z_2^m/z_2^s, z_3^m/z_3^s]$ or some of their logratios, i.e. with either one or the other we can estimate the subcomposition of the sample from the subcomposition of the standard, but not the absolute abundances of these 3 components.

Figure 1 shows that varying width of the measuring window or varying the sensitivity of one channel have the same influence on enrichment factors as the estimation of an average with varying sample size: the larger the total number of counts registered, the less variance the enrichment factor shows. On the other hand, varying the level of noise (i.e. the counts per unit of time of the background) does not affect the variability of results significantly, at least as long as the background represents less than 1:5 parts of the total signal. In any of the cases presented, estimates of the enrichment factors appear to be unbiased. The same can be said of the associated perturbations (results not shown).

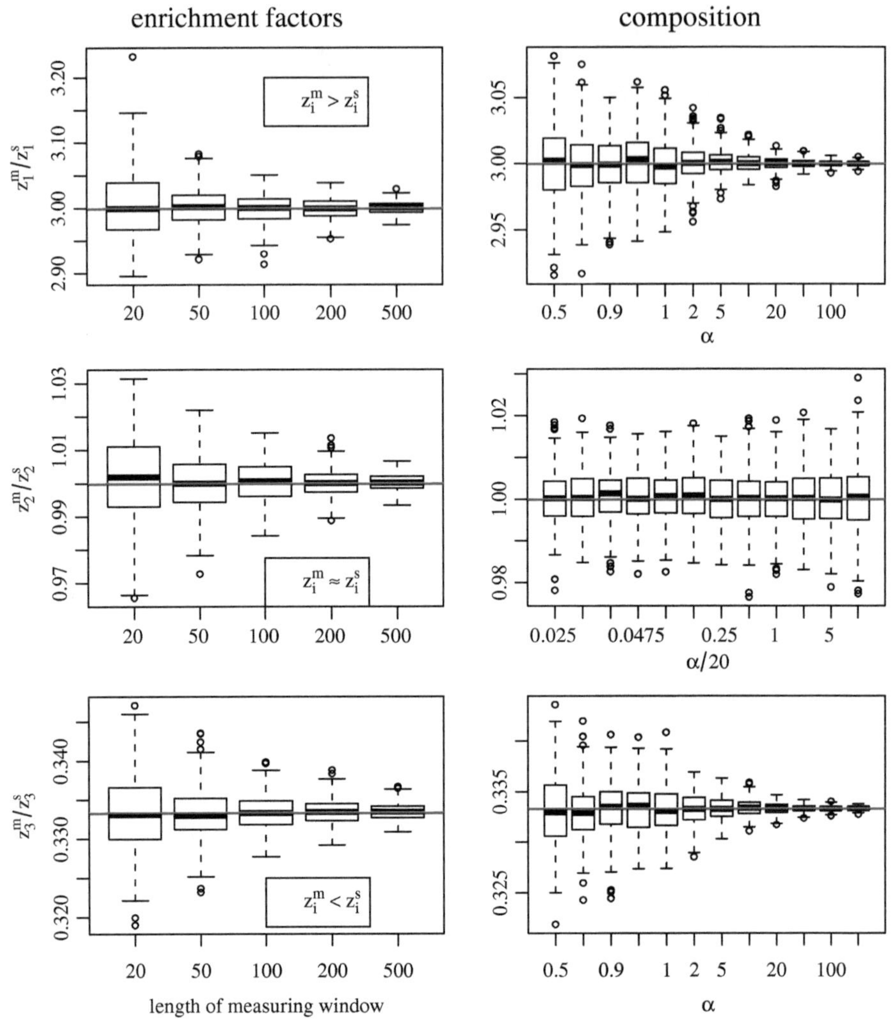

Fig. 1 Boxplots of enrichment factors obtained for: (*left*) varying lengths of the measuring window; (*right*) several values of α, with background level $\boldsymbol{\lambda}_b = [1, \alpha, \alpha]$ and sensitivity $\boldsymbol{\lambda} = 100[\alpha, 1, \alpha]$; in the *middle right* plot the X axis reports $\alpha/20$ which corresponds to the percentage of expected counts coming from the background compared with the total expected counts; on the contrary, the X axis upper and *lower right* plots directly reports α

A different picture is obtained if the ablated mass per unit of time of standard and of sample are not equal (Fig. 2). If the ablation rate is lower (higher) in the sample than in the standard, less (more) counts of all elements will be produced per unit of time, and the method will consequently estimate a lower (higher) absolute abundance of each element of the subcomposition. Absolute enrichment factors will therefore be strongly biased as long as the ratio of ablated masses differs from 1.

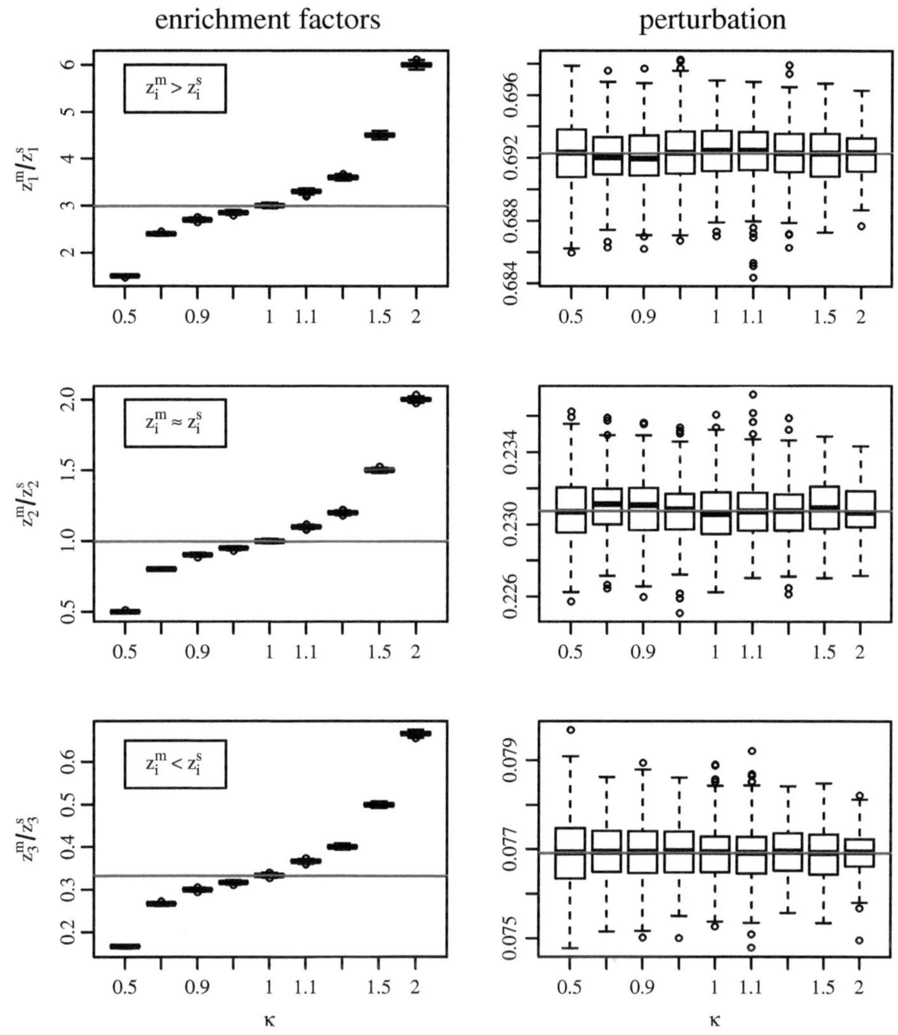

Fig. 2 Boxplots of enrichment factors and perturbations obtained for several ratios of ablated mass per unit of time from the sample and from the standard, denoted κ

On the contrary, perturbations remain fairly stable on a very wide range of this ratio (results shown between 1:2 and 2:1). Note that the reference levels are in this case $\mathcal{C}[3, 1, 1/3] = [9, 3, 1]/13$. These facts are well known to geochronologists, who typically work with ratios, but not so common in other geochemical communities using LA-ICP-MS data.

Figures 3 and 4 show the obtained enrichment factors in absolute or relative scales (as logratios) in the case that the real composition of the standard is assumed to vary

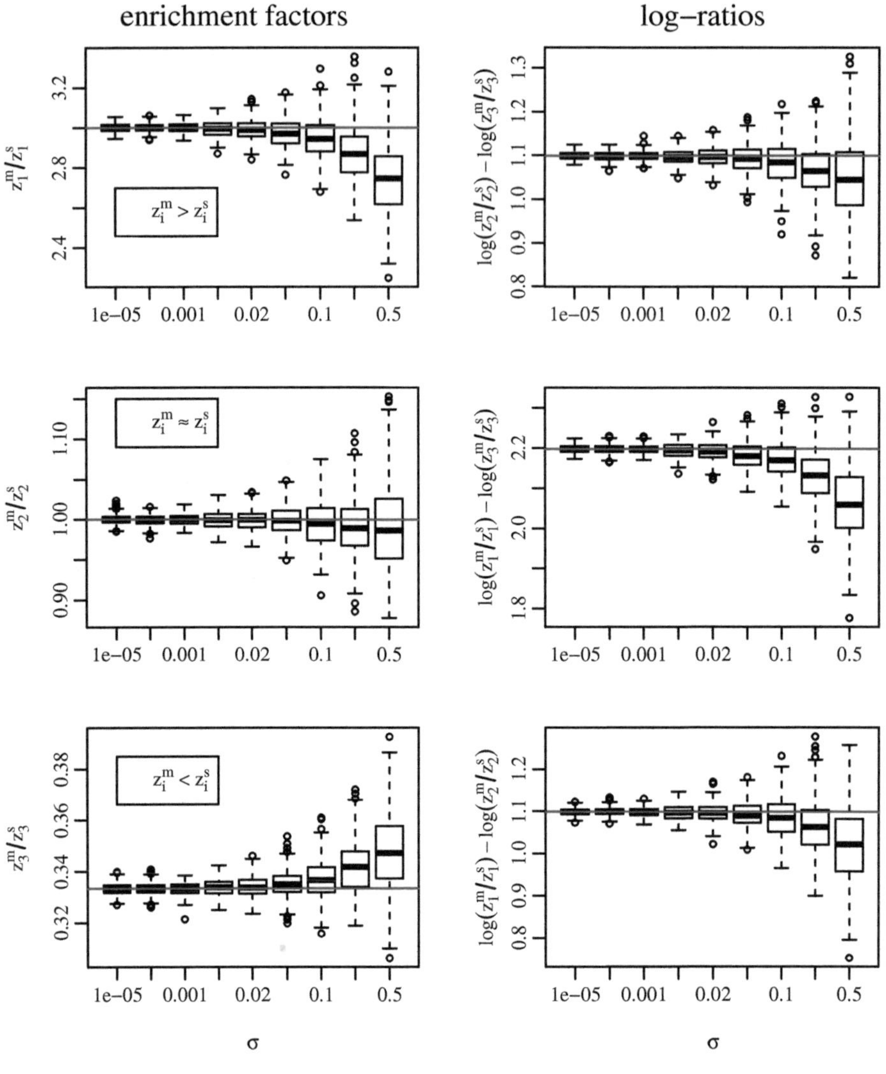

Fig. 3 Boxplots of enrichment factors and of their logratios obtained for several variability levels σ of the composition of the standard, which nominal value equal is taken as the closed geometric mean

along the spot, i.e. that masses with different compositions are sent to the detector during the measuring window of the standard. The actual composition is taken under assumption of additive logistic normal distribution with a diagonal ilr-variance matrix with variances σ within the subcomposition of the first three components, and 10^{-10} on the balance of the fourth component against the other three. For the parameters given, this distribution is undistinguishable from a univariate lognormal

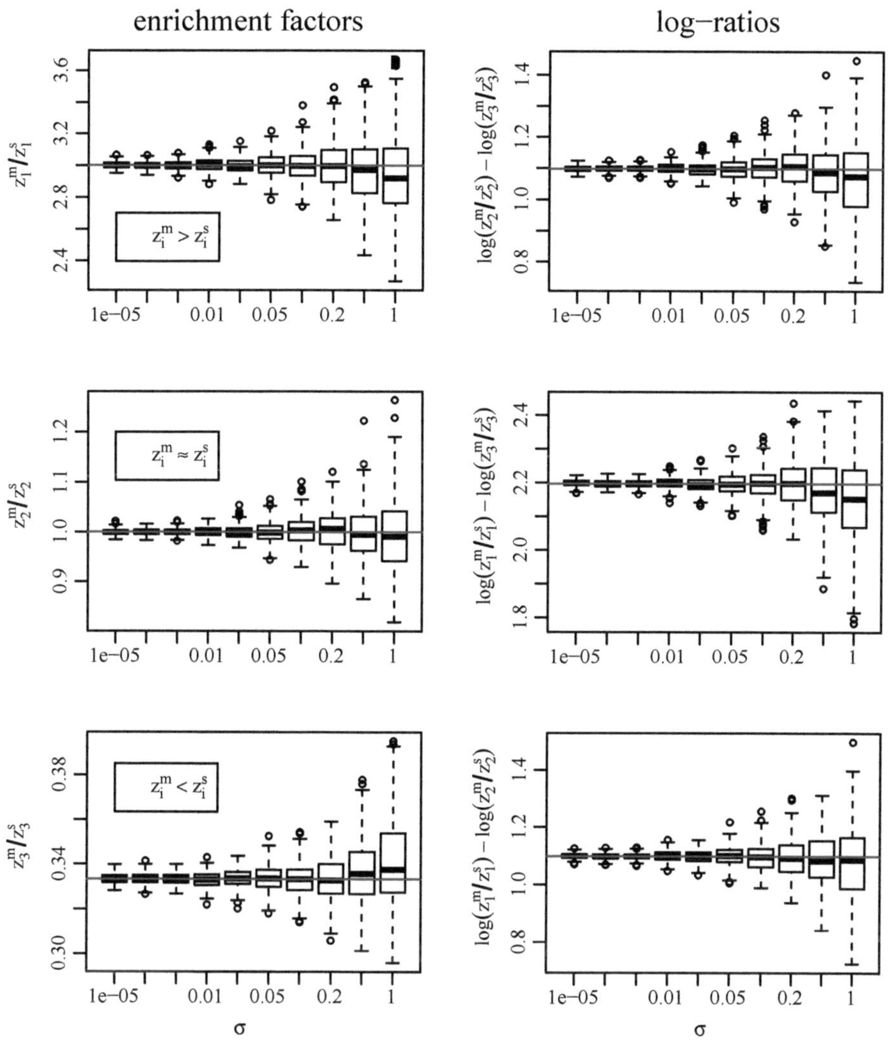

Fig. 4 Boxplots of enrichment factors and of their logratios obtained for several variability levels σ of the composition of the standard, which nominal value equal is taken as the arithmetic mean

distribution on each of the first three components. The two figures differ in which has been considered to be the nominal value: [1, 2, 3, 94] is taken as the compositional mean (closed *geometric* mean) of the standard in Fig. 3, while in Fig. 4 the *arithmetic* mean is forced to be the nominal value. Results show that a calibration equating nominal value to the geometric mean is not appropriate for moderate to high levels of standard heterogeneity (i.e. high values of σ) as both enrichment factors and logratios show notable biases in this case: enrichment factors larger than one tend to be underestimated, while those smaller than one tend to be overestimated. On the other hand, a calibration equating nominal value to the arithmetic mean shows almost no bias even in cases of high variance, though ultimatively the same overestimation/underestimation bias patterns occur. This highlights the importance of having homogeneous standards. Interestingly, this arithmetic specification of the nominal values delivers better (in the sense of less bias) results than geometric specifications even regarding the estimation of logratios.

Figure 5 (right) shows the results from the case that the ablated mass show an exponential decay fraction $\dot{m}(t) = \dot{m}_0 \exp(\ln(1 - \theta_t)t/\Delta t)$ of $100\theta_t$ % of the original mass during the time period Δt (which is considered equal to the measuring window

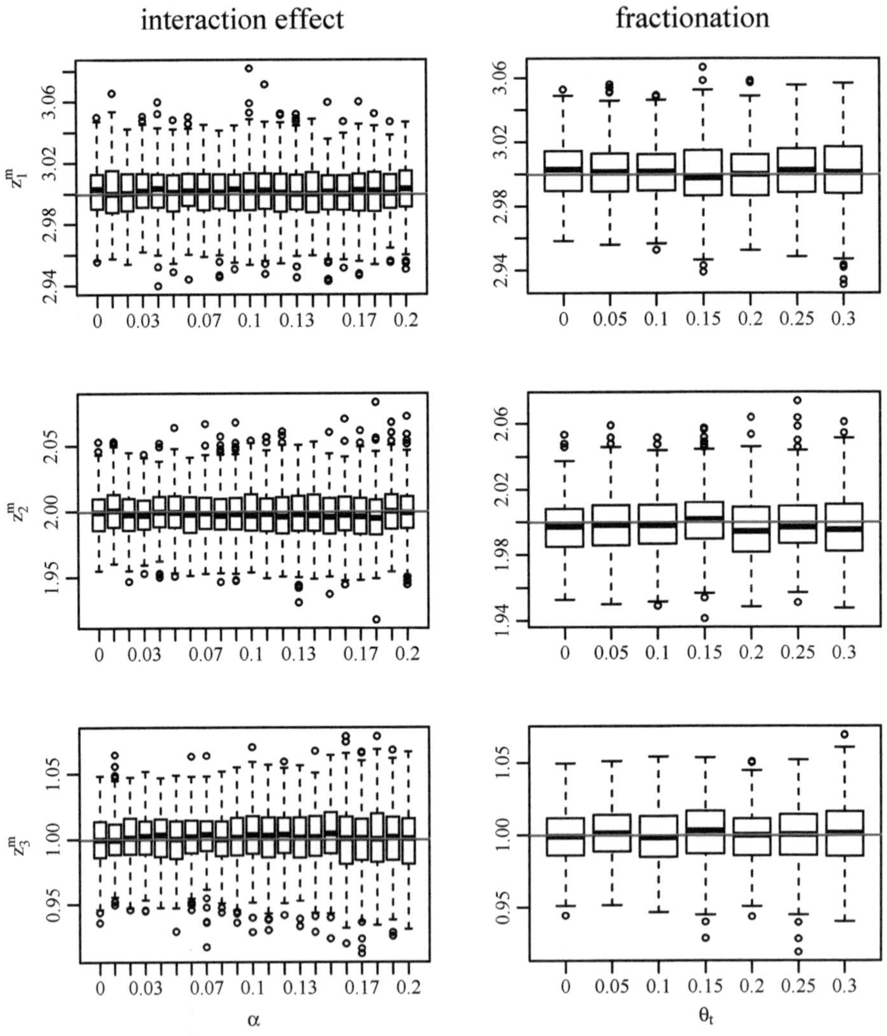

Fig. 5 Boxplots of estimated compositions obtained: (*left*) with different levels of interaction; (*right*) with different intensities of fractionation. See text for details

length). Note that the fitting GLM algorithm suggested cannot take this effect into account, thus we are dealing here with a model misspecification effect. In spite of this, final estimates show no degradation: the same variability and no bias can be observed, independently of the level of fractionation.

Finally, Fig. 5 (left) shows a case in which some interaction between components exist. Simulations are obtained with a constant matrix-interaction amount-perturbation model (Eq. 4), using the matrix

$$\boldsymbol{\Lambda}^* = \begin{pmatrix} 1 & \frac{\alpha}{1+\alpha} & \frac{\alpha}{1+2\alpha} \\ 0 & \frac{1}{1+\alpha} & \frac{\alpha}{1+2\alpha} \\ 0 & 0 & \frac{1}{1+2\alpha} \end{pmatrix}$$

for several values of α between $\alpha = 0$ (no interaction) to $\alpha = 0.2$ (notable interaction). In the calibration phase, the model parameters cannot be estimated with the classical setting (one single standard measured several times) because of collinearity. It is required to measure several standards: in this case, we have assumed nominal values in the subcomposition of interest $[1, 2, 3]$, $[1, 4, 9]$, $[9, 4, 1]$ and $[2, 3, 4]$, each of which was shot for a period of $N = 25$ readings. Moreover, the model fitted is rather the complete model, i.e. a matrix-interaction amount-perturbation model after Eq. (2). Results show no bias and a constant variability with increasing interaction parameter α.

In summary, the GLM methods proposed are applicable in the presence of moderate variability in the standard composition (the nominal composition should be the arithmetic expected value of the distribution of the composition of the standard), and do very well filter out exponentially decaying fractionation effects. However, if one cannot ensure that both sample and standard behave in the same way (similar masses per second ablated, i.e. the "matrices match") then only (log)ratios of the components are reasonably estimated.

6 Application

To illustrate the presented concepts, models and solving techniques we use a data set of geochronologically relevant isotopes. These are obtained ablating 35 zircons, including 9 standards, analysed for 6 isotopes: Hg202, Pb204, Pb206, Pb207, Pb208, Th232, U235, U238 (Fig. 6). The analysed sequences is composed of 3 blocks of 10 samples bracketed by 4 blocks of 5 standards (3 types of standards were used, coded GJ1, FC1 and 915, see Fig. 9). It is also possible to see that, except for some outliers, the isotopes are ordered in decreasing order of abundance as U238, then Th232 or Pb206, and finally U235, Pb208 and Pb207 in varying orders. The dwell times considered are reported in Table 1.

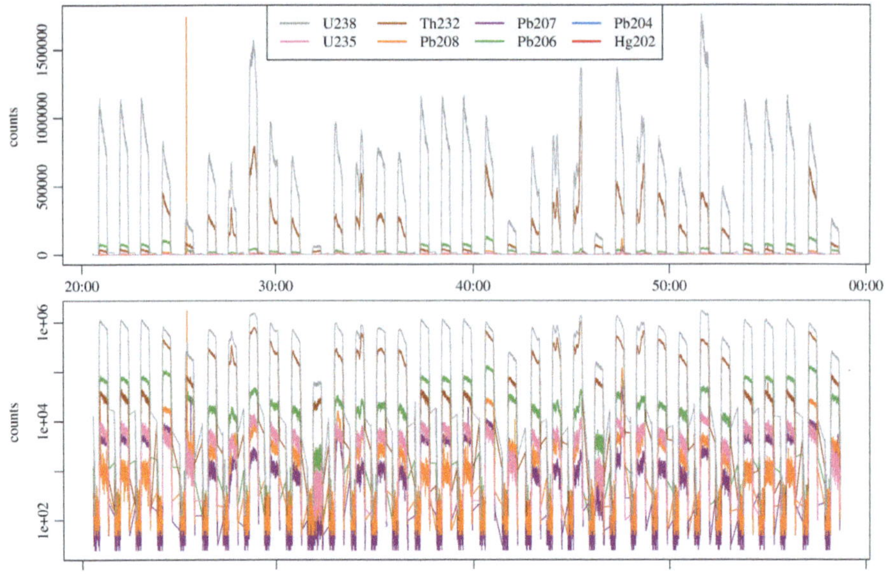

Fig. 6 Time series of counts obtained for each component considered, in raw scale (*top*) and in log scale (*bottom*), for 35 analytes. The *X* axis reports the time at which each measurement was obtained

Table 1 Dwell times for the several elements considered in ms

Hg202	Pb204	Pb206	Pb207	Pb208	Th232	U235	U238
4	4	2	8	4	2	2	2

A closer look at the last three measurement periods (Fig. 7), including one standard and two samples, shows several remarkable aspects:

- First, Hg202 and Pb204 do not show up in the samples or in the standard in a way significantly different from the background. These elements are used for analytical quality control, and are not relevant for the problem itself.
- Second, several isotopes show clear exponentially decreasing trends while being shot (linear in log scale), an effect of fractionation of the gas cloud while the laser penetrates deeper in the material.
- Finally, a more relevant aspect for modelling, the variability of these time series is neither additive nor multiplicative, because the background and the signal do not have comparable levels of variability neither in raw nor in log scales. This is the reason why purely multiplicative models (like those presented in Sect. A.2.1) or purely additive models (like using linear regression directly to link counts with composition, as done sometimes in the device calibration literature) are not realistic. This gives a diagnostic tool to decide whether a problem might be attacked with linear regression on raw variables (raw *Y*-axis plots show similar variability

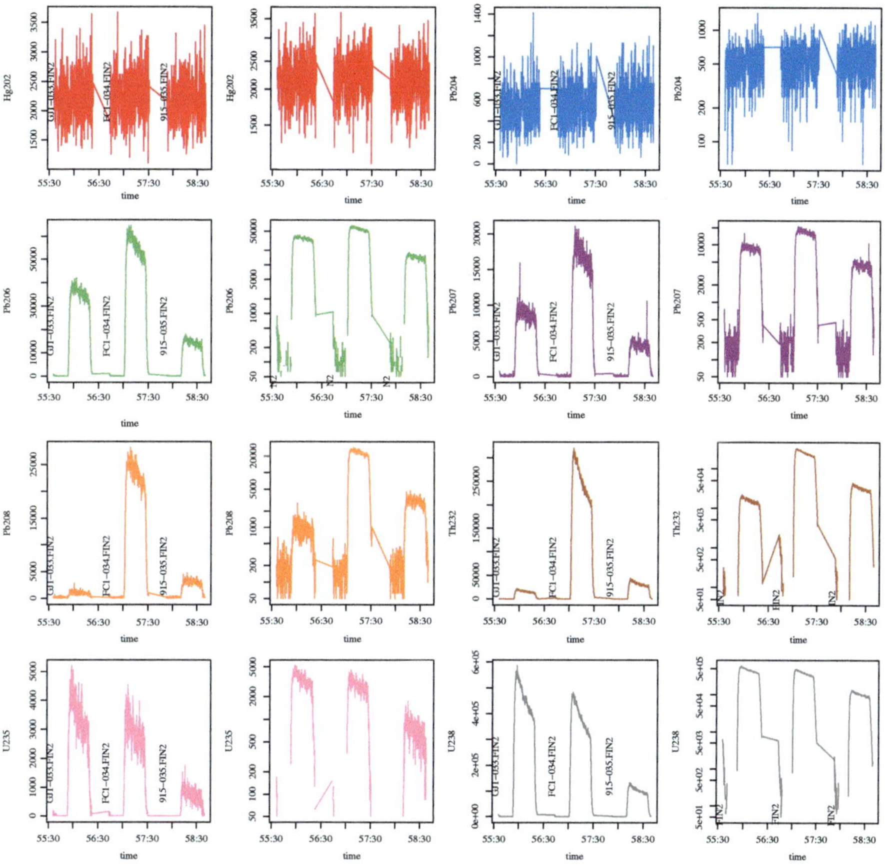

Fig. 7 Time series of counts obtained for each component considered, in raw scale and in log scale, for the last three samples. In particular, note that the variability of the data has neither a pure additive nor a pure multiplicative structure

in the background and signal windows), with linear regression on log-transformed variables (log Y-axis plots show similar variability in the background and signal windows), or Poisson regression (neither one nor the other are satisfied).

Other aspects and relevant concepts are shown as well in Fig. 8. Beside the clear fractionation effects occurring while the laser is shot (increasing trend, followed by a shallow exponential decrease) or right after switching it off (pronounced decrease), we see as well the presence of zonations of different composition, thus potentially of different age. The figure also shows that readings have much stronger drifts and variabilities than the relevant isotopic logratios in the measuring windows.

A further assessment on the possible structure of the variability of this series is displayed in Fig. 9, which makes use of the property that Poisson distributed variates should show dispersion coefficients of $\hat{D} = 1$. For the background windows,

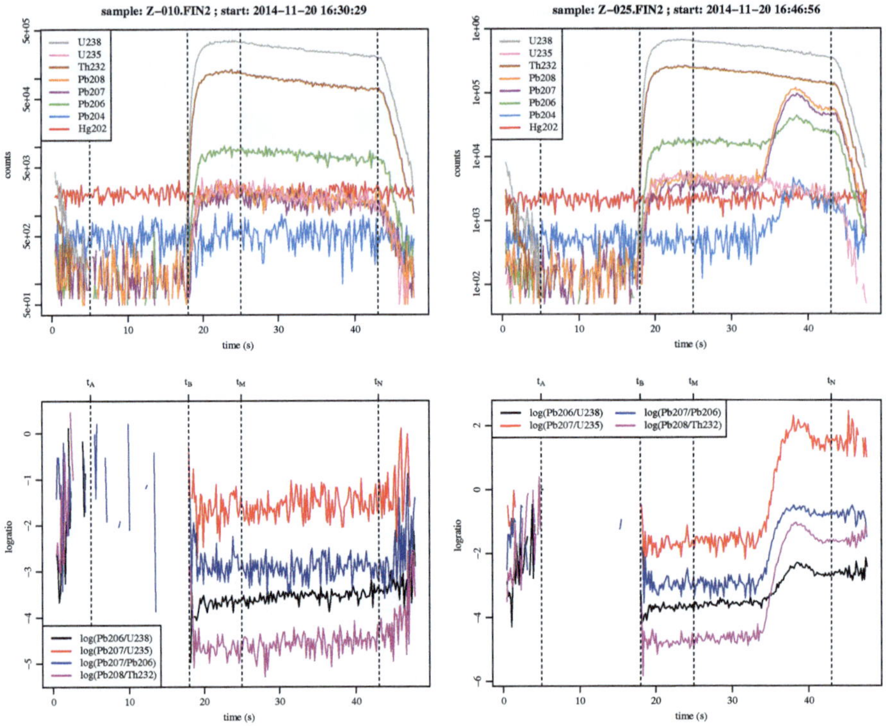

Fig. 8 Time series of counts obtained for each component considered, for a well-behaved sample (*left*) and for a sample showing clear zonation (*right*). *Upper plots* show the counts (in log scale) while *lower plots* show the naive logratios relevant for geochronological calculations. Note as well the *dashed vertical lines*, showing the windows for background (between the first two lines) and measurement (between the last two lines)

overdispersion is clear to see in the heavy ions (U238, Th232). Dispersion coefficients for the measurement period are only reported for the sake of completeness, as they are difficult to estimate due to the presence of the irregularities mentioned before: the values reported in this case are obtained using the residual variance with respect to a fitted exponential trend, which might deliver reasonable estimates for well-behaved samples, but is completely inappropriate for zoned samples. Nevertheless, it shows roughly a similar distribution with a certain bias towards underdispersion.

Given the considerations of this preliminary descriptive analysis, the natural conclusion must be that appropriate models for this dataset should be flexible enough to consider at least the following three effects:

1. natural variability on the composition of the materials (i.e., lack of homogeneity), in particular including zonation of samples and random inhomogeneities of standards and samples;
2. downhole fractionation (typically appearing as local exponentially decreasing trends);

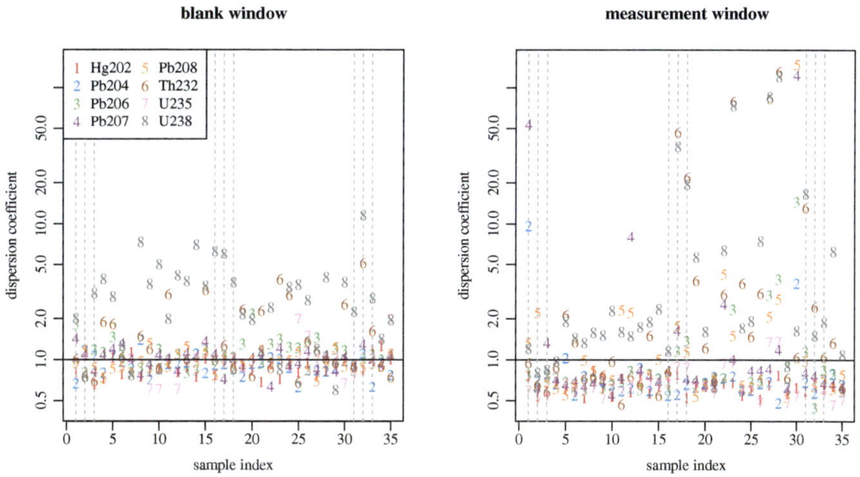

Fig. 9 Dispersion coefficients of the background and measurement windows for each sample. *Vertical dashed lines* mark the standards

3. Poisson overdispersion (something which is included in state-of-the-art GLM fitting models);
4. additive background variability.

Unfortunately, none of the models presently implemented within the framework of generalized linear models can deal with all these effects, the main limitation being that one cannot simultaneously treat as linear the additive background and the exponential trends. Nevertheless, as we have opted for leaving Bayesian methods for future work, the following is an approximate set of results ignoring the first and third shortcomings using a Poisson GLM regression with identity link (Sect. 4.2.1).

Results (Fig. 10) show several interesting patterns. The enrichment factors for the standards (Fig. 10 lower plot) should be all 1. The standards show a remarkably constant composition, suggesting that no long-range time drift is necessary. The method proposed is quite robust, for instance showing no to minor influences of the outliers of Pb208 in samples 915-005 or Z-025. If we order the elements by decreasing abundance on the estimated composition and on the observed counts, results are the same, which given the several shortcomings we had to take is a good result.

Finally, using constant nominal values for the standard GJ1 as reported in Table 2, the geochronologically relevant ratios (radiogenic ^{207}Pb versus ^{235}U, and radiogenic ^{206}Pb versus ^{238}U) of all 35 samples were computed. Given that no evidence of the presence of non-radiogenic ^{204}Pb is found (except perhaps in some zoned Zircons, like Z-025, Fig. 8), all ^{207}Pb and ^{206}Pb detected were considered to be radiogenic. Thus, the direct ratios ^{207}Pb/^{235}U and ^{2067}Pb/^{238}U were used to compute an age,

$$\frac{^{x}\mathrm{Pb}}{^{y}\mathrm{U}} + 1 = \exp(\lambda_y t),$$

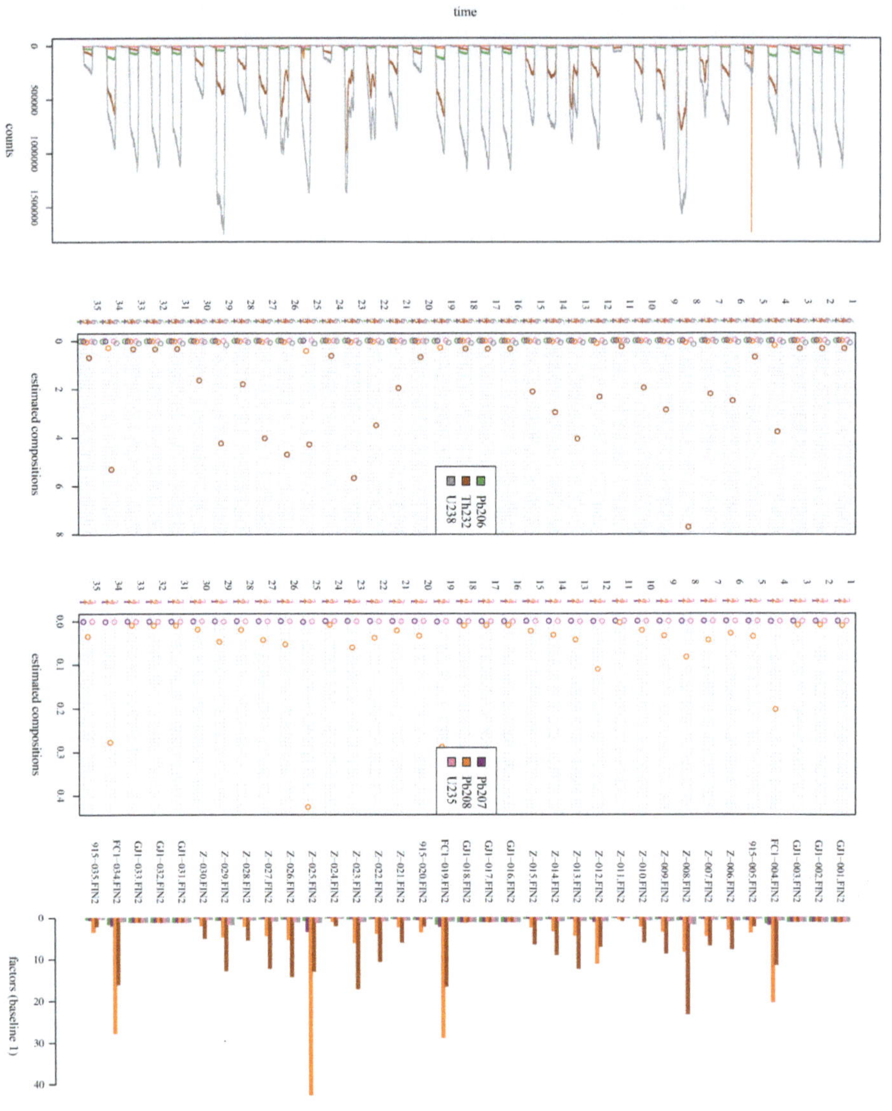

Fig. 10 Results of the analysis of counts for each isotope: (*1*) original data, (*2*) results for the major components, (*3*) results for the minor components, and (*4*) enrichment factors with respect to standard GJ1

Table 2 Nominal values of some isotopic ratios in standard GJ1 ()

	$^{206}Pb/^{238}U$	$^{207}Pb/^{235}U$	$^{208}Pb/^{232}Th$	$^{206}Pb/^{207}Pb$	Age (Ma)
GJ1	0.09761	0.8093	0.03011	0.06014	602

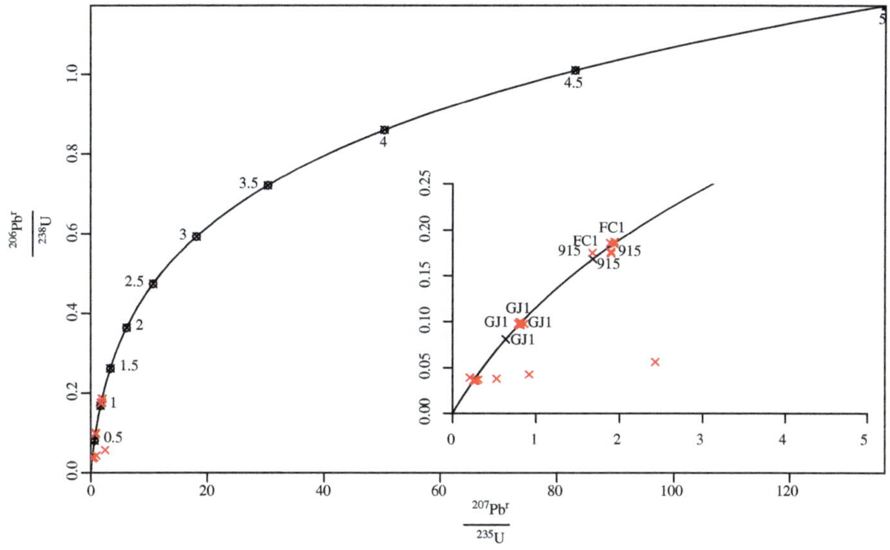

Fig. 11 Results of the geochronological ratios obtained for each

which for ^{207}Pb/^{235}U uses the decay rate $\lambda_{235} = 9.8485 \cdot 10^{-10}$, and for ^{206}Pb/^{238}U the rate $\lambda_{238} = 1.55125 \cdot 10^{-10}$. Figure 11 shows the so-called concordia diagram [14], a graphical display of the agreement between the two ages. The plot shows notable variability in the ages provided to the standards (especially, 915), and two samples that are notably far from the concordia curve, corresponding to zoned Zircons. But in general terms, the pair of ages show a satisfactory agreement for the whole data set. The obtention of some confidence regions on these calculated ratios within the scope of the GLMs used here remains to be done, and is left for future research.

7 Discussion

The most obvious implication of these results is the fact that models without additive error cannot be accepted as reasonable descriptions of the physical LA-ICP-MS measurement process. This is especially true when the target value is small because then the contribution of the multiplicative error to the measurement uncertainty becomes irrelevant in comparison with the contribution of the additive error. This was visible on the reduction of the estimated concentration of the minor isotopes of the standards in the real case study. This mixed additive-multiplicative nature could also be seen in the simulation studies with heterogeneous standard compositions: there an arithmetic mean specification of the standard delivered better results even with regard to the estimation of logratios. Though it is premature to extract conclusions of this single study, the additive nature of this kind of data perhaps comes from the fact that the ICP-MS does actually count individual atoms/ions/isotopes, and matter (or mass) is an additive property. Building a composition out of the obtained measurements of the several isotopes is then a choice of the analyst, and not intrinsically demanded by this kind of data.

A second family of implications relates to the fact that (log)ratios seem to be much more robustly estimated than their numerator and denominator elements separately, in particular if one cannot ensure that similar amounts of mass are ablated per second in the standard and the samples, i.e. if no good matrix matching is possible. This might be one of the reasons why many ICP-MS users do actually work with ratios, instead of with the absolute abundances provided by the device.

Third, results have implications in the treatment of values below the detection limit. Values below detection limit occur when the number of counts/sec. in the signal window cannot be statistically distinguished from counts/sec. in the background window, i.e. when the number of counts coming from the analysed mass is "low". In the authors' opinions, this strongly suggests to avoid additive logistic normal (ALN) models for BDLs, because the low number of counts does not allow the assumption of a central limit related normal approximation of the distribution. Moreover, the ALN totally disregards the additive effect of the background, which happens to be dominant precisely for the low values. An appropriate alternative should take somehow the counting nature of the problem into account.

Finally, to obtain a full joint compositional calibration model remains a difficult task because of the need of several standards of different composition, which should be each homogeneous, perfectly known and of the same kind of matrix than the samples to analyse. Given the practical problems of fulfilling these conditions in real-world applications even for a single standard, it is foreseeable that ICP-MS calibration will remain univariate. Nevertheless, the fact that perturbation enrichment factors (hence logratios) are notably robust to matrix mismatch opens the door to the possibility to of calibrating devices with several standards of relatively different matrices, and use the results for a full joint compositional calibration along the multi Poisson model shown here.

Appendix A

A.1 Three Competing Geometries

In this paper, three of the compositional geometries gathered by van den Boogaart and Tolosana-Delgado [12] are used, two on vectors of positive amounts ($\mathbf{x} \in \mathbb{R}_+^D$), either based on an interval or on a ratio scale; and a relative ratio scale on compositions *s.s.* ($\mathbf{x} \in \mathcal{S}^D$, the simplex). This section just summarizes their geometries.

The interval scale on $\mathbb{R}_+^D$ is captured by a geometry inherited from embedding $\mathbb{R}_+^D$ on $\mathbb{R}^D$ with its Euclidean space operations, namely the classical vector sum $+$

and multiplication by scalars $\cdot$. The fundamental isometric operation is the identity, thus this geometry is optimally reproduced by linear models with an identity link function.

The ratio scale on $\mathbb{R}_+^D$ is captured by equipping this set with the Abelian group operation *(amount)-perturbation* $\oplus_+$ and *(amount)-powering* $\odot_+$, respectively, the component-wise product of the components of two vectors and the component-wise powering of the components of one vector to the same scalar [6]. The fundamental isometric operation is the component-wise log-transformation, thus this geometry is optimally reproduced by linear models with a logarithmic link function.

The relative scale on $\mathcal{S}^D$ is captured by equipping this set with the Abelian group operation *perturbation* $\oplus_+$ and *powering* $\odot_+$, respectively the closed component-wise product of the components of two vectors [1] and the component-wise powering of the components of one vector to the same scalar. The fundamental isometric operation is the centered log-ratio transformation, thus this geometry is optimally reproduced by linear models with a logratio link function.

In general terms, a raw scale should be preferred for one variable which absolute differences are meaningful; a ratio scale for variable which meaningful differences are relative; and a relative scale on the simplex should be the choice if several variables show a ratio scale at the same time and their sum is either an artifact or a meaningless constant. Nevertheless, sometimes the same variables can be studied in one way or another, depending on the question that should be answered, i.e. the scale should be chosen depending on the question and not only on the data.

A.2 Alternative Joint Models

A.2.1 Models Ignoring the Background or With Multiplicative Effects

From the point of view of the relative scale on $\mathcal{S}^D$, all models presented in this paper are particularly complicated by the presence of the additive background. If this could be ignored (e.g. because it is very small with regard to the signal), and the sum of the components of $\mathbf{Z}$ equals one (i.e. a whole composition is analysed), then the following models are derived:

- scalar upscaling: $[\mathbf{X}(t_k)|n] \sim \mathcal{M}u(\mathbf{Z}; n)$ and $n \sim \mathcal{P}o(\omega_0 \lambda(t_k))$;
- perturbation: $[\mathbf{X}(t_k)|n] \sim \mathcal{M}u(\mathbf{Z} \oplus \boldsymbol{\lambda}(t_k); n)$ and $n \sim \mathcal{P}o(\omega_0 \sum_i^D \lambda_i(t_k))$,;
- interaction-perturbation: $[\mathbf{X}(t_k)|n] \sim \mathcal{M}u(\mathcal{C}\left[\boldsymbol{\Lambda}^* \cdot (\mathbf{Z} \oplus \boldsymbol{\lambda}(t_k))\right]; n)$ and $n \sim \mathcal{P}o$ $(\omega_0 \mathbf{1}' \cdot (\boldsymbol{\Lambda}^* \cdot (\mathbf{Z} \oplus_+ \boldsymbol{\lambda}(t_k))))$.

This last model is still a mixture of additive and multiplicative geometries. The following models are purely compositional alternatives, using in the compositional part only operations on the simplex

- scalar upscaling: $[\mathbf{X}(t_k)|n] \sim \mathcal{M}u(\mathbf{Z} \oplus \boldsymbol{\lambda}_b; n)$ and $n \sim \mathcal{P}o(\omega_0 \lambda(t_k) \sum_i^D \lambda_{bi})$;
- perturbation: $[\mathbf{X}(t_k)|n] \sim \mathcal{M}u(\mathbf{Z} \oplus \boldsymbol{\lambda}(t_k) \oplus \boldsymbol{\lambda}_b; n)$ and $n \sim \mathcal{P}o(\omega_0 \sum_i^D \lambda_{bi} \lambda_i(t_k))$;

- interaction-perturbation: $[\mathbf{X}(t_k)|n] \sim \mathcal{M}u(\mathbf{\Lambda}^* \boxdot (\mathbf{Z} \oplus \boldsymbol{\lambda}(t_k)) \oplus \boldsymbol{\lambda}_b; n)$ and $n \sim \mathcal{P}o(\omega_0 \mathbf{1}' \cdot (\mathbf{\Lambda}^* \boxdot (\mathbf{Z} \oplus \boldsymbol{\lambda}(t_k)) \oplus \boldsymbol{\lambda}_b))$, in this case with $\oplus$ the perturbation on the simplex.

In these expressions, we have used the notation $\boxdot$ after Pawlowsky-Glahn, Egozcue and Tolosana-Delgado ([7], Chap. 4) to denote a simplicial endomorphism operation, i.e. one such that once expressed in any basis of the simplex becomes a simple matrix-vector product. Note that these purely compositional models imply, among other effects, that the noise induced by the background upscales with the signal, i.e. larger signal should show more background variability.

In any of the cases presented, we finally have distributions for the number of counts on each element class that belong to the multi-Poisson family, with an intensity vector model $\omega_0 \mathbf{\Lambda}(\mathbf{Z}, \boldsymbol{\lambda}_b, \boldsymbol{\theta}; t_k)$ capturing the relationship between the expected partial counts and the composition of the analyte. Hence, we can always consider that the total number of counts $n(t_k)$ follows a Poisson distribution with $\lambda_k^T = \omega_0 \mathbf{1}' \cdot \mathbf{\Lambda}(\mathbf{Z}, \boldsymbol{\lambda}_b, \boldsymbol{\theta}; t_k)$; and, conditional on that total, the vector of counts for each element follows a multinomial distribution with probability parameter vector $\mathbf{p}_k = \mathcal{C}\left[\mathbf{\Lambda}(\mathbf{Z}, \boldsymbol{\lambda}_b, \boldsymbol{\theta}; t_k)\right]$.

A relevant minor modification for the examples presented in this paper (Sects. 5 and 6) consists of the case that the dwell times are not equal for all isotopes. If we denote the vector of dwell times as $\boldsymbol{\omega}_0 \in \mathbb{R}_+^D$, then it is immediate to show that the number of counts on each element class belong to the multi-Poisson family, with a perturbed intensity vector model $\boldsymbol{\omega}_0 \oplus_+ \mathbf{\Lambda}(\mathbf{Z}, \boldsymbol{\lambda}_b, \boldsymbol{\theta}; t_k)$, with perturbation on $\mathbb{R}_+^D$. Again, the total number of counts $n(t_k)$ will follow a Poisson distribution, albeit with total expected counts $\lambda_k^T = \boldsymbol{\omega}_0' \cdot \mathbf{\Lambda}(\mathbf{Z}, \boldsymbol{\lambda}_b, \boldsymbol{\theta}; t_k)$; and the vector of counts for each element conditionally follows a multinomial distribution with probability parameter vector $\mathbf{p}_k = \mathcal{C}\left[\boldsymbol{\omega}_0 \oplus_+ \mathbf{\Lambda}(\mathbf{Z}, \boldsymbol{\lambda}_b, \boldsymbol{\theta}; t_k)\right]$.

A.2.2 Implications for the Estimation of GLMs

The methods presented in the main part of this contribution were characterized by an identity link function, a requirement of the additive nature of the background and the signal. If this condition is removed, then the class of models can be extended to models with logarithmic link function (actually, the canonical choice of Poisson GLMs). This setting would be suitable to consider the multiplicative models mentioned in Sect. A.2.1. The same structure as in Sect. 4 would then be used, namely a calibration phase in which all parameters would be estimated with a GLM; and a prediction phase in which another GLM with an offset (equal to the multiplicative background) would be used to estimate the unknown composition of the samples.

References

1. Aitchison, J.: The Statistical Analysis of Compositional Data (Reprinted in 2003 by The Blackburn Press), p. 416. Chapman & Hall Ltd., London (UK) (1986)
2. Cheatham, M.M., Sangrey, W.F., White, W.M.: Sources of error in external calibration ICP-MS analysis of geological samples and an improved non-linear drift correction procedure. Spectrochim. Acta **48B**(3), E467–E506 (1993)
3. Donevska, S., Fišerová, E., Hron, K.: Calibration of compositional measurements. In: Communications in Statistics—Theory and Methods, accepted (2016). doi:10.1080/03610926.2013.839039
4. Jackson, S., Pearson, N., Griffin, W., Belousova, E.: The application of laser ablation-inductively coupled plasma-mass spectrometry to in situ U–Pb zircon geochronology. Chem. Geol. **211**, 47–69 (2004)
5. Nelder, J., Wedderburn, R.: Generalized linear models. J. R. Stat. Soc. Ser. A **135**(3), 370–384 (1972)
6. Pawlowsky-Glahn, V., Egozcue, J.J.: Geometric approach to statistical analysis on the simplex. Stochast. Environ. Res. Risk Asses. **15**(5), 384–398 (2001)
7. Pawlowsky-Glahn, V., Egozcue, J.J., Tolosana-Delgado, R.: Modelling and Analysis of Compositional Data. Wiley, Chichester (2015)
8. Skellam, J.G.: The frequency distribution of the difference between two Poisson variates belonging to different populations. J. R. Stat. Soc. Ser. A **109**(3), 296 (1946)
9. Sylvester, P.J.: Matrix effects in laser ablation-ICP-MS. In: Sylvester, P.J. (ed.) Laser Ablation ICP-MS in the Earth Sciences: Current Practices and Outstanding Issues, pp. 67–78. Mineralogical Association of Canada, Short Course 40 (2008)
10. van den Boogaart, K.G., Tolosana-Delgado, R., Hron, K., Templ, M., Filzmoser, P.: Compositional Regression with unobserved components or below detection limit values. In: CoDa-Work'13, 5th international workshop on Compositional Data Analysis, held on 4–7 June 2013, Vorau (Austria)
11. van den Boogaart, K.G., Tolosana-Delgado, R.: Analyzing compositional data with R. UseR series, 258p. Springer, Hiedelberg (Germany) (2013)
12. van den Boogaart, K.G., Tolosana-Delgado, R.: "Compositions": a unified R package to analyze compositional data. Comput. Geosci. **34**(4), 320–338 (2008)
13. van den Boogaart, K.G., Tolosana-Delgado, R., Templ, M., Filzmoser, P.: Regression with compositional response having unobserved components or below detection limit values. Stat. Model **15**(2), 191–213 (2015)
14. Wetherhill, G.: Discordant uranium-lead ages, I, transactions. J. Am. Geophys. Union **37**, 320–326 (1953)

MIX
Papier aus verantwortungsvollen Quellen
Paper from responsible sources
FSC® C105338

If you have any concerns about our products,
you can contact us on
ProductSafety@springernature.com

In case Publisher is established outside the EU,
the EU authorized representative is:
Springer Nature Customer Service Center GmbH
Europaplatz 3, 69115 Heidelberg, Germany

Printed by Libri Plureos GmbH
in Hamburg, Germany